AF352928

Sustainable Energy Systems: Efficiency and Optimization

Sustainable Energy Systems: Efficiency and Optimization

Editors

Alan Brent
Toshihiko Nakata

MDPI • Basel • Beijing • Wuhan • Barcelona • Belgrade • Manchester • Tokyo • Cluj • Tianjin

Editors
Alan Brent
Sustainable Energy Systems
Victoria University of Wellington
Wellington
New Zealand

Toshihiko Nakata
Graduate School of Engineering
Tohoku University
Sendai
Japan

Editorial Office
MDPI
St. Alban-Anlage 66
4052 Basel, Switzerland

This is a reprint of articles from the Special Issue published online in the open access journal *Energies* (ISSN 1996-1073) (available at: www.mdpi.com/journal/energies/special_issues/ Sustainable_Energy_Systems_Efficiency_and_Optimization).

For citation purposes, cite each article independently as indicated on the article page online and as indicated below:

LastName, A.A.; LastName, B.B.; LastName, C.C. Article Title. *Journal Name* **Year**, *Volume Number*, Page Range.

ISBN 978-3-0365-1586-1 (Hbk)
ISBN 978-3-0365-1585-4 (PDF)

Contents

About the Editors

Alan Brent

Alan Brent is the inaugural Chair in Sustainable Energy Systems at Victoria University of Wellington, New Zealand. The Chair supports the transition to sustainable energy in New Zealand as well as in the regional economy and society. This is aligned with the strategic sustainability focus at Victoria University of Wellington, to meet current and future challenges, by taking an inter-disciplinary approach, and engaging with partners across society in trans-disciplinary ways.

Toshihiko Nakata

Toshihiko Nakata heads the Energy Systems Analysis Laboratory in the Graduate School of Engineering of Tohoku University, Japan. His laboratory focuses on the analysis and integrated design of sustainable energy systems.

Preface to "Sustainable Energy Systems: Efficiency and Optimization"

New, alternative energy technologies are rapidly becoming affordable, and it is expected that these will be immensely disruptive to our traditional mode of centralised energy generation, transmission, and distribution. Additionally, the severe climate (and other) impacts of many traditional energy generation and utilisation techniques are widely accepted. As stated in the World Energy Outlook of the International Energy Agency (IEA), "the global energy scene is in a state of flux, thrown off balance by falling costs for a range of technologies, led by wind and solar...". Furthermore, the IEA states that it "...expects renewable electricity generation to increase by a further two-fifths by 2021. However, renewable heat and transport are lagging behind, despite good potential...". For these changes or transitions to be just and sustainable, systemic analyses are required, with an emphasis on optimizing the energy sector to be better integrated with the other sectors of the economy, thereby ensuring the efficiency of future energy value chains. Such systemic analyses utilise concepts, methods and tools such as system dynamics, urban metabolism, industrial ecology, and life cycle analyses, to inform policy- and decision-making.

This book is a collection of papers from a Special Issue in the journal *Energies*, with contributions from researchers across the globe. The book provides insights to scholars on how the concepts, methods and tools of systemic analyses have been utilised in various contexts—from the household and community level, to industrial and utility-scale levels—to enable a transition to sustainable energy systems, with an emphasis on the efficiency and optimisation of future energy value chains.

Alan Brent, Toshihiko Nakata
Editors

Article

Conceptualizing Household Energy Metabolism: A Methodological Contribution

Adél Strydom [1], Josephine Kaviti Musango [1,*] and Paul K. Currie [1,2]

[1] School of Public Leadership, Stellenbosch University, Stellenbosch 7600, South Africa; adelstrydm@gmail.com (A.S.); paul.currie@iclei.org (P.K.C.)
[2] ICLEI Africa, Cape Town 7441, South Africa
* Correspondence: jmusango@sun.ac.za

Received: 3 June 2019; Accepted: 22 August 2019; Published: 29 October 2019

Abstract: Urban metabolism assessments enable the quantification of resource flows, which is useful for finding intervention points for sustainability. At a household level, energy metabolism assessments can reveal intervention points to reshape household energy consumption and inform decision-makers about a more sustainable urban energy system. However, a gap in the current urban metabolism research reveals that existing household energy consumption studies focus on outflows in the form of greenhouse gas emissions, and have been mostly undertaken at the city or national level. To address this gap, this study developed a method to assess household energy metabolism focusing on direct energy inflows in the form of carriers, and through-flows in the form of services, to identify intervention points for sustainability. Then, this method was applied to assess the energy metabolism of different households in Cape Town, South Africa, as categorized by income groups. The study argued that the developed method is useful for undertaking bottom–up household energy metabolic assessments in both formal and informal city settings in which more than one energy carrier is used. In cities where only national or city-level data exists, it provides a method for understanding how different households consume different energy carriers differently.

Keywords: urban metabolism; household energy metabolism; household energy metabolism assessment methods; household energy consumption

---

1. Introduction

Energy is an integral part of human well-being, and is one of the basic services that humans require in order to thrive. It provides the means to cook, to heat, to cool, to light up homes, to charge mobile phones, and to operate other electronic devices. The modern energy system causes significant environmental impacts. Most (60–80%) of the global final energy consumption is in urban areas, and is responsible for between 71–86% of greenhouse gas emissions [1,2], making it a large contributor to climate change. Given the central role that cities play as energy consumers, it is crucial to understand the different energy consumption patterns therein so as to identify intervention points for reshaping energy flows toward a more sustainable energy system.

Cities can be examined at different levels, of which the household is the smallest structural level [3]. As a sector, the household is crucial driver of energy consumption in cities [4–7]. Despite technical innovation in the form of energy efficiency and renewable energy technologies, household energy consumption continues to rise [8,9]. Therefore, households are a critical point of intervention.

Many households in the Global South lack access to high-quality energy services [10]. In particular, inadequate electricity access results in these households consuming alternative energy carriers that are often inefficient or physically harmful. These households are expected to experience an increase in energy consumption, which means increased greenhouse gas emissions [11]. Hence, the focus in the

Global South should be on improving access in the most efficient way while addressing high levels of consumption in other parts of the city [12], ultimately reducing greenhouse gas emissions while increasing access.

Urban metabolism uses the concept of flows to understand how resources move through a city [12], making it possible to understand flows that shape or have the potential to reshape urban areas to become sustainable. Thomson and Newman [13] highlighted that a city's metabolism consists of many metabolisms. Therefore, it is vital to take a deeper exploration into cities to explore metabolisms of different resource types or metabolisms at different levels. For instance, it can be scaled to address specific resources (e.g., energy) at specific levels (e.g., household).

Studying energy at the household level provides the closest look at how human activity contributes to energy consumption. It follows that researchers persistently recommend a household focus for metabolism studies [14–16]. Although various urban metabolism tools have been developed, including Integrated Land Use, Transportation, Environment (ILUTE) (ILUTE "simulates the evolution of an integrated urban system over an extended period of time "[17]) [17], Integrated Urban Metabolism Analysis Tool (IUMAT) [18], and URBANISM ("*UrbanSim* is an integrated transportation land-use model" [19]) [19], few practical assessments have been done on energy metabolism at the household level.

It is essential to understand how household characteristics and activities shape energy flows. This includes addressing not only electricity, but also alternative energy carriers servicing many households in the Global South, such as charcoal and paraffin. While some households may consume large amounts of energy and have the potential for reducing this consumption, others still require sufficient access to energy carriers that do not threaten their health. Existing household energy metabolism assessments tend to use disaggregated city-level data, resulting in coarse estimates. In light of this, bottom–up assessments may offer robust insights on how different household types use energy. Further, both direct and indirect household energy consumption are overwhelmingly examined in the literature. However, this paper focuses only on the direct energy consumption, because (i) it is specifically focused on energy access, which is a concept that refers only to direct energy, and (ii) indirect energy is the subject of further research, particularly around implications of transportation as well as water and food consumption.

After presenting an overview of the literature related to urban energy metabolism and household energy consumption, the objective of this paper is therefore to present a method developed for undertaking a differential household energy metabolism assessment, focusing specifically on inflows in the form of energy carriers and through-flows in the form of energy services. Further, the method is applicable to both formal and informal settings within the urban environment, where formal dwellings include brick and mortar houses and apartments in high-income areas, and where informal dwellings are often built from scrap building materials and located in low-income areas. The method was applied to the city of Cape Town, South Africa, as it is a city where bottom–up household energy data are severely lacking. The city is also home to a broad range of income groups and dwelling types, making it a useful location for developing and applying the method.

2. Urban Energy Metabolism for Sustainability

The origin of the term 'urban metabolism' is highly contested. In 1883, Marx first imagined the notion that a society as a whole has a resource metabolism, in which nature is transformed as needed to provide society with the necessary commodities [11,20,21]. Some argue the first explicit mention of the term 'urban metabolism' was made in 1965 by Wolman, who presented the metabolism of a hypothetical American city to demonstrate the metabolic needs of a city, such as the materials the city's inhabitants need to sustain their home, work and leisure lives [8,16,20,22]. Meanwhile, others believe that Theodor Weyl pioneered the term in 1894, in his discussion of food consumption, comparing nutrient discharge with the food intake into the city of Berlin [23].

The most frequently cited definition of urban metabolism is that of Kennedy et al. (2007, p. 44), which defines it as "the sum total of the technical and socioeconomic processes that occur in cities, resulting in growth, production of energy, and elimination of waste". The authors originate from the Industrial Ecology discipline and, as the definition suggests, are particularly focused on the quantification of resource flows. The definition is criticized for being too restrictive in its implied methods and practical application of an urban metabolism assessment [3,11]. The main criticism is the bias toward quantification and consequent disregard for the emergent properties that are possible through resource exchange [11]. Similarly, Barrera et al. [3] argued that the definition should also include the social and political aspects of a city, such as how resources are distributed. Currie and Musango [14] and Musango et al. [11] call for the inclusion of people and information flows, while acknowledging the role of built and natural systems in conveying flows.

Currie and Musango [14] thus define urban metabolism as the "collection of complex sociotechnical and socioecological processes by which flows of material, energy, people, and information shape the city, service the needs of its populace, and impact the surrounding hinterland". This definition includes a significantly broader scope, namely a shift from a purely accounting view to one that accounts for complexity and an explicit consideration for the needs of the residents in a city, thereby addressing not only efficiency, but also equity. However, this paper aims to understand a very specific aspect of resource flows in the city: that of energy consumption within households. For this specific application, the broad definition provides a foundation of accounting for various technical flows into and out of the city. This foundation aids in contextualizing flows of specific resources on a specific scale of the city, which, in this, case is household energy.

While the theoretical approach is developed enough for practical applications, urban metabolism assessments of cities are still limited, particularly in the Global South [8,11,21]. Possible reasons are that: (i) urban metabolism assessments lack standardized methods [11,16,24] and (ii) are simpler to conduct where rich data for resource flows already exist [25]. Another argument is that cities have different contexts, which implies that a standard method might not be as viable to large-scale practical applications of urban metabolism as the literature suggests.

Urban metabolism can highlight intervention points for lightening resource dependence in cities. Its inclusion of all types of resources may hinder its ability to make practical and spatially explicit recommendations for cities. This paper is of the similar view with Carréon and Worrell [20] that in order to progress toward a sustainable energy system, an important starting point is to understand the flows of energy through the city. Approaching a sustainable energy system also means understanding how people access and use energy differently in the city, to ensure access to safe, reliable, and modern energy sources to all citizens.

The Urban Energy System

Energy is a unique resource to examine, as it does not flow in the same manner as most resources. Instead, it flows through the different phases of the urban energy system. Zhang et al. [26] disaggregate the system into five phases: namely energy exploitation, energy transformation, industry, living, and recovery. Carreón and Worrell [20] indicate three phases of energy system, namely (i) energy sources, which are connected by (ii) energy carriers to meet (iii) the city's energy demand. Both studies emphasize that energy flows from one phase to the next. Therefore, understanding the energy system requires understanding urban energy flows.

A city's energy system can be conceptualized as depicted in Figure 1, which was adapted from Carreón and Worrell [20], Chen and Chen [27] and Zhang et al. [28]. Energy exploitation is the first phase of the system, which allows the identification of the source of the various energy flows. This phase includes all the mining activities for raw materials. In the second phase, energy is transformed into carriers. The physical infrastructure of grids, refineries, and power plants transform energy into the carriers of fuel, electricity, gas etc., which hold the energy.

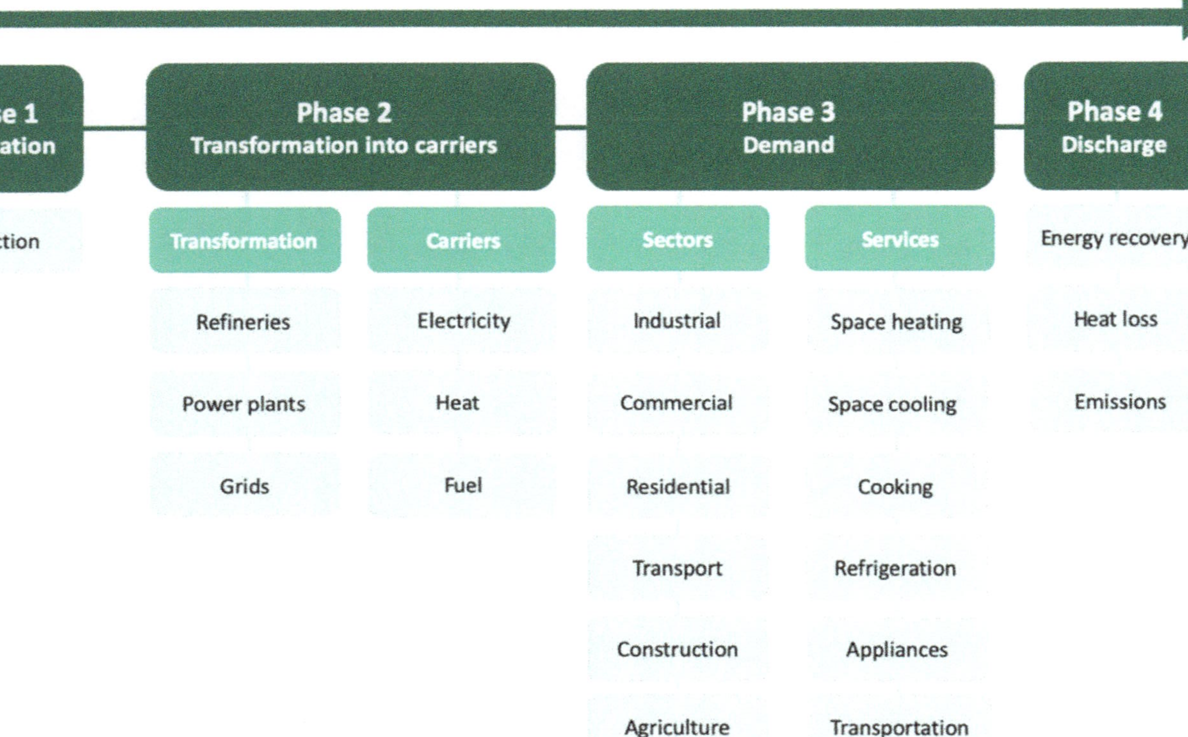

Figure 1. The urban energy system. Sources: Adapted from Carreón and Worrell [20], Chen and Chen [27], and Zhang et al. [28].

The third phase is energy demand. According to Carreón and Worrell [20], this phase can be divided into energy sectors and end use, which represent two flows within the same phase. Figure 1 shows a possible further disaggregation in sectors based on Chen and Chen [27], who provided detail on the various energy sectors found in a city. Some disagreement exists around the difference between energy end use and energy services. What Carreón and Worrell [20] regarded as end use, Bristow and Kennedy [29] and Barrera et al. [3] regarded as services. Carreón and Worrell [20] regarded energy services as a further phase after energy end use. This study makes use of the views of Bristow and Kennedy [29] and Barrera et al. [3] based on Fell [30], who regarded energy services as the function performed using energy. A fourth phase needs to be added in order to account for Zhang et al.'s [28] energy recovery. The top arrow in Figure 1 indicates that the entire system represents the flow of energy from extraction to processes in the city, and finally discharge into the environment. Not all energy may flow through all phases, but it always flows from left to right, from exploitation to discharge.

3. Towards Conceptualizing Household Energy Metabolism

Various urban energy metabolism assessments have been conducted. Chen and Chen [27] translated the city's energy activities into carbon flows in order to model the carbon metabolism and associated energy-use activities. Zhang et al. [31] studied the energy metabolism of various sectors in a city as well as their associated carbon footprints. Both studies argue that carbon flows should be central to urban energy metabolism assessments, as this helps to understand the carbon profile of cities and consequently the amount of pressure that a city's energy system places on the environment and consequently the contribution of a city to climate change risk [27,31]. Therefore, the focus is mainly on the energy outflows. Weisz and Steinberger [32] similarly focus on the energy outflows in their review on the various ways in which a city can reduce both its energy and material flows. The notion to focus strongly on greenhouse gas emissions is limited. Energy assessments also require expansion beyond carbon to include the local dynamics of energy provision and use, so as to understand future energy demand, infrastructure pressures, and how to effectively plan fast-growing cities.

There is another prevailing gap in addressing the through-flows of energy within the urban system. While Chen and Chen [27] deemed the flows between sectors important, they mainly addressed the inflows in the form of extraction and the eventual outflows to the carbon sink. This correlates with the first and last phase of the energy system, and leaves a gap for addressing the flows within, namely carriers, sectors, and services. This gap is addressed by Zhang et al. [28], who argued that urban metabolism struggles to address ecological trophic levels within the energy system. They shifted the focus to analyze the relationships within this system using through-flow analysis and ecological network utility analysis, and found a total of 73 different metabolic pathways between 17 energy sectors, concluding that it is possible to make the city's energy flows more efficient by adjusting these relationships [28]. For example, to balance out a system where demand is higher than supply, energy consumption must either be lowered, or energy production must be increased. Consequently, it is possible to grasp the adequacy of supply of primary energy sources to meet the needs of the energy service phase.

Carreón and Worrell [20] identified another gap: most energy metabolism studies are overwhelmingly linear, using only accounting approaches and input–output analysis, disregarding causal relationships between elements such as climate, demographics, and infrastructure. Beyond quantification are possibilities to address the other mentioned aspects of sustainability, such as equity. The energy literature is firmly embedded in a perspective of reducing and controlling. Weisz and Steinberger [32] argued that energy access is widely overlooked in energy metabolism studies. Therefore, interventions in areas that still lack access to energy might involve increasing inflows or shifting the energy carrier within a certain flow in order to provide more reliable, affordable, or efficient energy services. The concept of a sustainable energy system should go beyond quantifying carbon emissions and reducing energy flows [33]. Brunner's [34] call to reshape flows rather than to reduce or control once again arises, and the energy metabolism literature can benefit from this perspective, as reshaping will allow for increased flows or a change in energy carriers where necessary.

Another gap in the literature on urban energy metabolism is that most of these studies were done at the city level, taking a top–down approach using national data, thereby disregarding the dynamics of space and time within a city [20]. This study addresses this gap by focusing on household energy metabolism, given the calls in Musango et al. [11], for conducting bottom-up research in cities in order to account for spatio-temporal dynamics.

Cities exist in various structural and societal scales and levels [3,11,35]. Barrera et al. [3] provided a useful representation of the levels within the energy system, distinguishing between the macro, meso, aggregated, and micro levels. Each of these levels can then be divided into behavioral and structural categories. Table 1 provides more detail.

Table 1. Levels of organization of energy systems.

Levels of Organization	Behavioral	Structural
Micro	Households, firms	Buildings (houses)
Aggregated	Urban land uses	Squares or neighborhoods; groups of buildings
Meso	Economic sectors	Urban districts
Macro	Economic sectors Cities	Cities

Source: Barrera et al. [3].

As Table 1 shows, the smallest scale of the energy system can be understood as buildings, or houses. On the behavioral level, it specifies the household as the smallest scale, which indicates a difference between a house and a household. However, the smallest scale of a city is the individual. Moll et al. [36] and Biesiot and Noorman [37] argued that while individuals perform different consumer activities, these are mostly focused within the household, and therefore the household, not the individual, is the smallest unit. In terms of energy consumption, this paper views the household as the smallest unit, as energy consumption within the household contributes to services that are shared between the individuals within. Furthermore, the household is a standardized unit in metabolic study, and the majority of energy consumption studies present their data for the household as a whole.

There exists no clear definition of a household energy metabolism. However, the literature provides definitions of household metabolism, from which a definition for energy in particular can be inferred. The most basic definition of household metabolism is "the integral patterns of natural resources flowing into and out of households" [37]. Donato et al. [38] provided further detail by defining household metabolism as the biophysical assessment of households from the point of view of raw materials, energy carriers, and water required, and emissions and wastes resulting from household consumption patterns. The inputs are further categorized in direct inputs of energy (electricity, heat, and vehicle fluids) and material, and indirect inputs of economic goods and services.

Based on the above, a household energy metabolism can be understood as the process by which energy flows are sourced and delivered through various carriers, which are conveyed through the house to service a household's direct and indirect energy requirements, and result in waste or emissions. Through-flows are equally important to inflows and outflows, justifying the reference to through-flows in the household energy metabolism definition.

A brief distinction between direct and indirect energy consumption in the household is needed, as total household energy requirements include both. Direct energy is related to the energy consumed within the household and includes energy for space heating, water heating, cooking, lighting, and electronics, while indirect energy is used for the production, transportation, and disposal of goods and services consumed by the inhabitants of the house [5,7,36]. While Benders et al. [7] indicate that the majority of household energy consumption research includes only direct energy, the case for including indirect energy has strengthened significantly, and other scholars have considered the total household energy to include indirect consumption [3,36,38].

Existing household metabolism assessments are mostly embedded in a sustainable development approach. However, they vary considerably with regard to the resources studied. Both Moll et al. [36] and Donato et al. [38] identified intervention points for making the household more sustainable, but the former addresses total household energy requirements, while the latter reviewed a body of household metabolism research papers, including both energy and material resources. Yang et al. [39] and Biesiot and Noorman [37] focused on the environmental effects of household resource consumption with a strong focus on greenhouse gas emissions. The former analyzes emissions from energy, material, food, and waste, while the latter is interested only in total energy consumption. Frostell et al. [40] also studied total household energy requirements, but went beyond accounting to find ways of changing the energy consumption behavior. Within this variance exists a strong focus to measure the emissions impact of households, whether this pertains to the energy consumption alone or a broader study of household resource requirements [38].

In their literature review, Donato et al. [38] argued that the methods for conducting household metabolism assessments have not yet reached maturity, and only provided a review of its status. When considering the diversity in approaches to the household's resources, it is understandable that most of the studies reviewed use hybrid methods. Moll et al. [36] and Biesiot and Noorman [37] used a combination of input–output analysis and process analysis to account for the complex nature of quantifying indirect energy consumption. Biesiot and Noorman [37] outlined that the direct energy requirements can be determined by: (i) considering the money spent on energy, (ii) dividing this into energy activities, (iii) accounting for the energy requirements of these activities, and (iv) converting this energy into CO_2 emissions. The framework is useful, as it provides a method to quantify various household activities in energy terms. The framework can be applied to different scenarios, thereby projected future energy consumption.

Top–down approaches dominate in household metabolism assessments. Biesiot and Noorman [37] and Yang et al. [39] acquired their data sets from national data and disaggregated it to the household level. Yang et al. [39] acquired supplementary data from household surveys. Both Moll et al. [36] and Donato et al. [38] argued for the need to undertake a bottom–up approach to assessing household metabolism. Moll et al. [36] found significant variances in consumption patterns between the countries they studied, and concluded that it is crucial to study different types of households before identifying intervention points. Biesiot and Noorman [37] shared a similar view that different households display different lifestyles, and therefore different consumption patterns. Therefore, a differential household energy metabolism approach is crucial in order to account of variances in consumption patterns, lifestyles, and countries or areas within specific cities.

In order to conduct a differential household energy metabolism assessment, reliable and accurate data is crucial [36]. Biesiot and Noorman [37] recommended collecting the following data sets before attempting a household energy metabolism assessment: (1) energy production and consumption data, (2) economic input–output matrices, (3) household budget surveys, and (4) goods and services price information.

The studies focusing specifically on direct household energy consumption fall mostly outside of the urban metabolism field, but prove useful in identifying methods for collecting bottom–up data. Smart meters in households are proving to be a very effective and reliable way of collecting quantitative energy consumption data, as they account for exact consumption [41–43]. Studies that do not utilize smart meters often measure electricity consumption using utility bills [15,44]. Neither one of these approaches are appropriate for a study on direct energy consumption across a range of income groups and energy carriers, as an approach is needed that accounts for the different energy sources that a household may access, which is not depicted in smart meter data. While smart meters are excellent for tracking direct electricity consumption, they do not indicate exactly how this electricity is used in the house—for example, which appliances it services. A holistic understanding of household energy consumption means examining how multiple carriers feed into a variety of services accessed.

Existing household energy metabolism assessments account for the total energy requirements and total emissions. Limited studies have explored how energy flows into and through the household [37,45]. To study energy through-flows, the concept of energy services is useful. Both Sovacool [45] and Barrera et al. [3] advocated for household energy consumption studies to address services. Barrera et al. [3] indicate that the usefulness of viewing energy flows as services is due to the simplicity in translating the energy activities performed in a household, thereby making explicit what the individuals in a household choose to consume. It relates energy consumption to activities, which is easier for consumers to comprehend than referring to the amounts of Joules, kilowatt-hours, or liters of carriers consumed. In this way, intervention points become more tangible or accessible to the individual.

The main energy services within the household across different regions are similar. Based on work in the United States of America (USA) and Western Europe, Abrahamse et al. [4] developed an energy service hierarchy and suggested that space heating is the highest energy consuming service, followed by water heating, refrigeration, lighting, cooking, and finally space cooling. Sovacool [45] examined middle-income households from a broad range of countries and suggested that the primary energy services (in order) are space heating, water heating, cooking, appliances, and lighting. Kwak et al. [46] studies North Korea, and found that space heating and space cooling were major contributors to energy consumption due to the country's four distinct seasons. In Finland, which experiences colder weather than most countries, space heating is the primary energy service [47].

In contrast to hierarchies developed by country, studies that compare energy services between low-income and high-income households have observed differing energy hierarchies. Sovacool [45] found that the energy in low-income households predominantly service lighting and cooking, while other surveys also include hot water, television, and radios. A study on energy services in rural African regions listed cooking, lighting, and water heating as primary energy services [10]. Offering an interesting contrast is the types of additional energy services found in high-income households: swimming in heated swimming pools or cooking with the television on are common findings [45]. This once again stresses the need for a differential household energy metabolism assessment in order to understand whether households in the same city may appear strikingly different when their energy inflows and through-flows are analyzed.

Sovacool [45] stressed that services also make it possible to identify the level of access of the household by a proposing an energy ladder. This energy ladder differs from the traditional energy ladder, which focuses only on carriers, and suggests that households transition to more efficient fuel types as their economic situation improves [48]. Sovacool's [45] energy ladder includes three drivers to be applied to the various steps of the ladder: satisfying subsistence needs; convenience, comfort and cleanliness; and conspicuous consumption [45].

The basic energy carriers for a large number of households around the world are fairly consistent. Sovacool [45] identified electricity, natural gas, coal, liquefied petroleum gas, kerosene, and fuel oil, with electricity being the most dominant energy carrier. However, as with differences in service between low-income and high-income households, subsistence households may demonstrate the widest range of energy carriers. Figure 2 conceptualizes the energy ladder and household energy services of various income groups according to Sovacool [45]. Households driven by subsistence only can be regarded as not having sufficient access to energy.

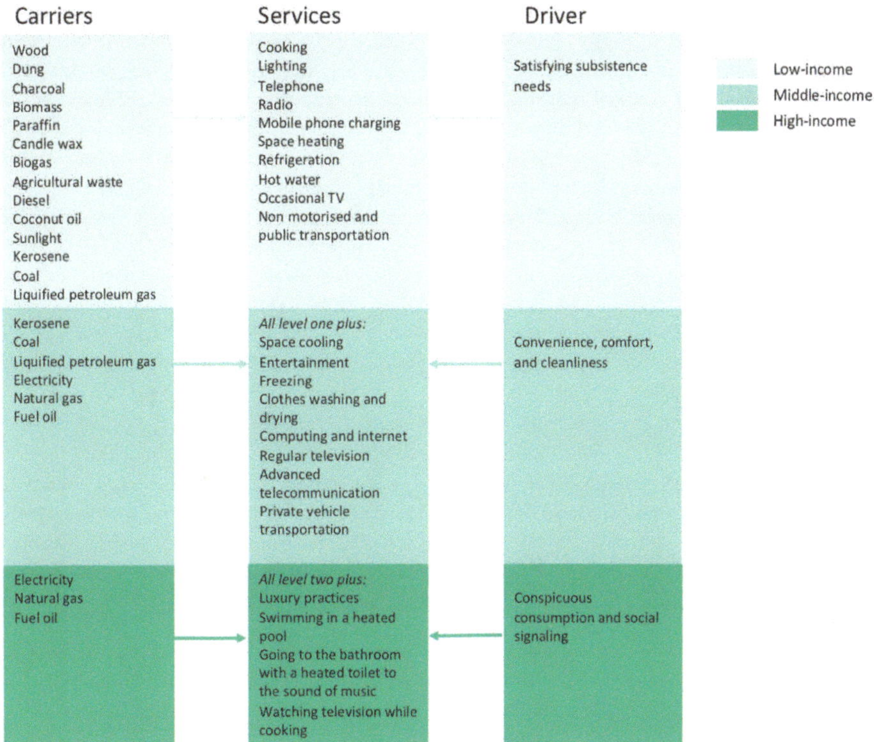

Figure 2. Household energy ladder, adapted. Source: Adapted from Sovacool [45].

A useful way to look at energy access is to consider the concepts of 'energy poverty', 'fuel poverty', and 'energy vulnerability'. All of these are conceptualizations that aim at identifying the group lowest on the energy ladder, but they differ quite significantly. Energy poverty is a term that is typically used to refer to inadequate energy access in the Global South, and links to the wider relationship between energy and development [49]. Fuel poverty and energy vulnerability refer to people typically in the Global North who have access to energy, but cannot afford to purchase sufficient amounts [49–51]. The difference between the two is that energy poverty is a state of being, while energy vulnerability can change according to external factors such as the dwelling quality, energy costs, and stability of household income [51]. Bouzarovski and Petrova [49] categorize all three terms under the umbrella term of 'domestic energy deprivation'.

It is essential to consider energy in terms of drivers rather than a state of being, as it provides a clear pathway for improving energy security, which is important especially in the Global South, where energy access is a concern. Rather than containing the household in a negative state of being, the energy ladder creates a conceptualization of being able to climb up to a position in which energy consumption ceases being driven by subsistence and starts being driven by comfort, cleanliness, and convenience.

To achieve a more sustainable energy system may in some cases mean changing the energy carrier. Camara et al. [52] explicitly stated the importance of addressing the forms in which low-income households access energy, as a change in energy carrier could result in higher energy efficiency. Therefore, these households could climb to the second level of the energy ladder without experiencing an increase in energy spending. Examples of key intervention points in the Global South are improved cook stoves and cleaner fuel, such as a transition away from solid fuels and paraffin toward gas or electricity and improved wood burning stoves with, for example, chimneys [10,53–57]. This shows that improved access to modern energy carriers (such as electricity) is not the only possible intervention, but that in some cases, changing the fuel type from solid fuel to liquid petroleum gas or kerosene (widely considered to be modern fuel types) can also be beneficial [54]. Improved cook stoves and cleaner fuel is mostly discussed within the context of improving health; however, Williams et al. [57] and Maes and Verbist [55] discussed it in conjunction with air pollution, and Williams et al. [57] included hardships experienced by the women collecting solid fuels on foot. Given that a sustainable energy system must consider both social and environmental factors, the quality of cooking fuel or technology must necessarily reduce pollutants, as well as negative health impacts.

This paper proposes a method for examining the relationship between carriers, services, and drivers of household energy consumption. It forms part of a larger research project, which includes household consumption of food and water, and production of wastes, and a number of methods were performed together with co-researchers. However, this paper retains focus on the methods that relate only to household energy metabolism.

4. Materials and Methods

In order to develop a method for conducting a differential household energy metabolism assessment, this paper created two key conceptualizations. The first is that of Cape Town's household energy flows in terms of carriers, services, and drivers, and the second illustrates the relationship between carriers and services in Cape Town households. In order to achieve this, various methods were used. Figure 3 illustrates the process and methods used to reach these conceptualizations.

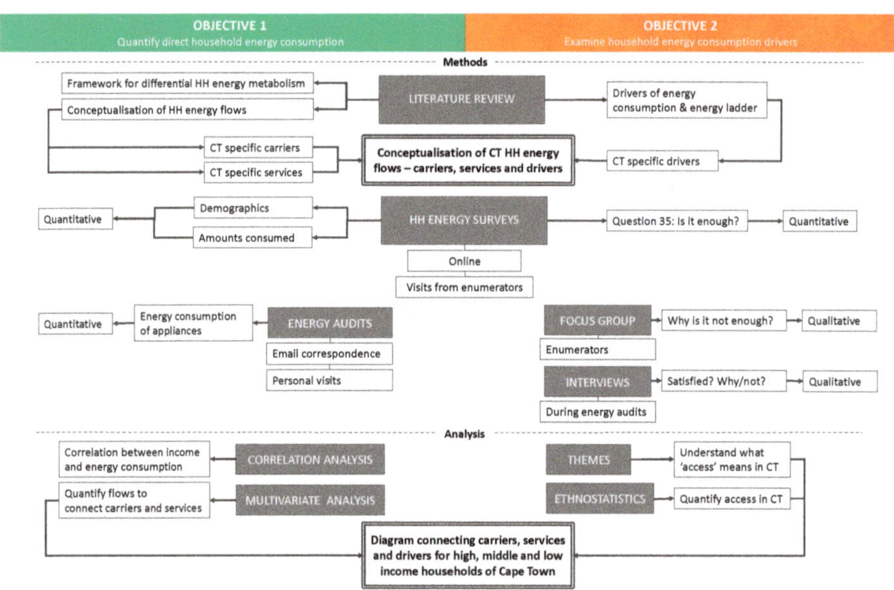

Figure 3. Method conceptualization. Note: CT is Cape Town; HH is household.

Defining components of the energy system was essential in conceptualizing Cape Town's household energy flows in terms of carriers, services, and drivers. However, since this energy system was created based on studies from the Global North, it was essential to adapt the carriers and services to correspond to the carriers and services accessed by the residents of Cape Town. This conceptualization is depicted in Figure 4. Based on the iterative process between the literature and the data collected, the paper made several changes to the services found in the literature:

- The literature distinguishes between refrigeration and freezing, but this paper categorizes the two together as Refrigeration, as the majority of respondents own combination fridge/freezers, therefore requiring a single classification.
- The service of Personal grooming was added, as appliances such as hair dryers and electric shavers are a regular addition to households in Cape Town.
- The literature refers to 'mobile phone charging'; however, as residents of Cape Town often use tablets for communication, mobile phones and tablets were grouped together as Communications.
- The service of House cleaning was added to account for the presence of dishwashers.

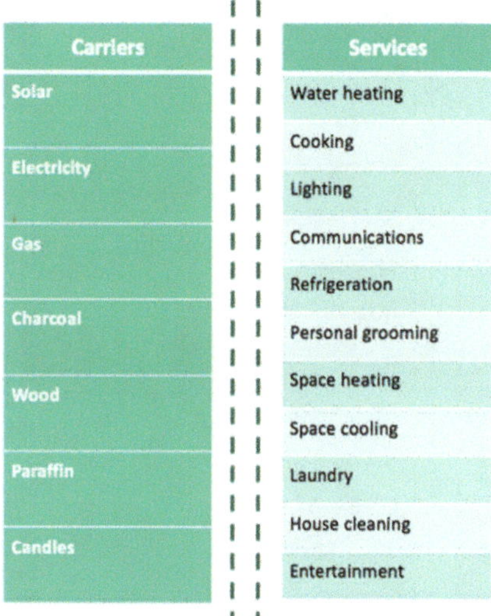

Figure 4. Cape Town household energy carriers and services.

The services of computing and entertainment were grouped together as Entertainment, since many residents watch television or on-demand streaming services on their computers.

Following Sovacool [45], it was important to identify the drivers that are specific to the study. Since this paper's focus was in households in Cape Town that still require sufficient access to energy, it was interested in the two drivers lowest on the energy ladder: how a household could move from 'satisfying subsistence needs' (not having sufficient energy access) to 'convenience, comfort, and cleanliness' (having sufficient access). For this reason, the top driver, namely 'conspicuous consumption', was omitted, as this focuses on the next level of the energy ladder (mostly associated with high-income households), and requires an in-depth study of its own.

Once the two energy drivers were selected, it was possible to add this information to the conceptualization of Cape Town's household energy flows in order to create a conceptualization of Cape Town's household energy flows—carriers, services, and drivers. This conceptualization is depicted in Figure 5. The services are reorganized according to the energy ladder in the literature, as well as the energy services and carriers applicable to Cape Town. The energy ladder specifies that low-income households are driven by satisfying subsistence needs, while middle-income households are driven by convenience, comfort, and cleanliness.

In order to understand the relationship between energy carriers and services, and thereby create the second conceptualization for Cape Town's household energy flows, a household energy survey with questions about the use of 44 key household appliances was designed and distributed in Cape Town. The appliances were categorized twice: first by energy carrier, and second by the energy services detailed in Figure 5. After answering basic questions around household size, household income, and dwelling type, respondents were asked to explain the frequency with which they use the 44 appliances. The answers were presented in brackets: for example, 2–3 h per day or 4–5 times per week. This allowed respondents to answer the surveys without having the exact amounts and times on hand. While this is not as accurate as a metering exercise, it allows the opportunity to understand how energy is consumed in households that use more energy carriers than electricity (for which metering exercises are ideal). This survey was also designed in such a way to be easily understandable to all residents and to gain a broad idea of how many households consume energy, rather than a specific idea of a small sample of households.

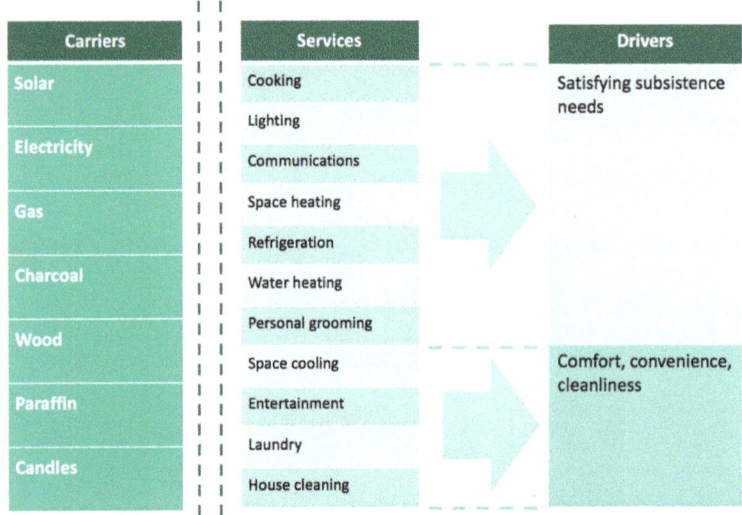

Figure 5. Conceptualization of Cape Town's household energy flows.

An online version of the survey was aimed at reaching middle-income and high-income households. A group of enumerators visited low-income suburbs of Cape Town with hard copies of the surveys. The income groups were determined using the local income groups, as reflected in the South Africa Statistics.

The total number of households surveyed was 676, with only 391 surveys that were completed in full. Only 360 surveys that were useful for Cape Town were utilized in this paper. Responses came in from 56 of Cape Town's 190 suburbs, with the most responses (66) received from the suburb of Khayelitsha. The enumerators surveyed a larger group of respondents than the online survey. However, some of their respondents earn middle or high incomes. This indicates that within suburbs such as Khayelitsha, Mfuleni, and Mitchell's Plain reside many middle-income and high-income residents, and that enumerators visited these households, too.

5. Results

Figure 6 represents energy consumption based on activity for the household brackets using the average consumption per bracket. This consumption is divided into energy carriers (the left-hand side of each diagram) and energy services (the right-hand side of each diagram). Energy carriers indicate the total amount of activity for each carrier for each income bracket, while energy services indicate the end-use of energy for each bracket. This diagram provides a basic understanding of how the relationship between carriers and services changes with income. Figure 6a provides insight on the household level, and Figure 6b shows energy consumption per capita, which highlights the effect on average consumption as household size decreases and income increases.

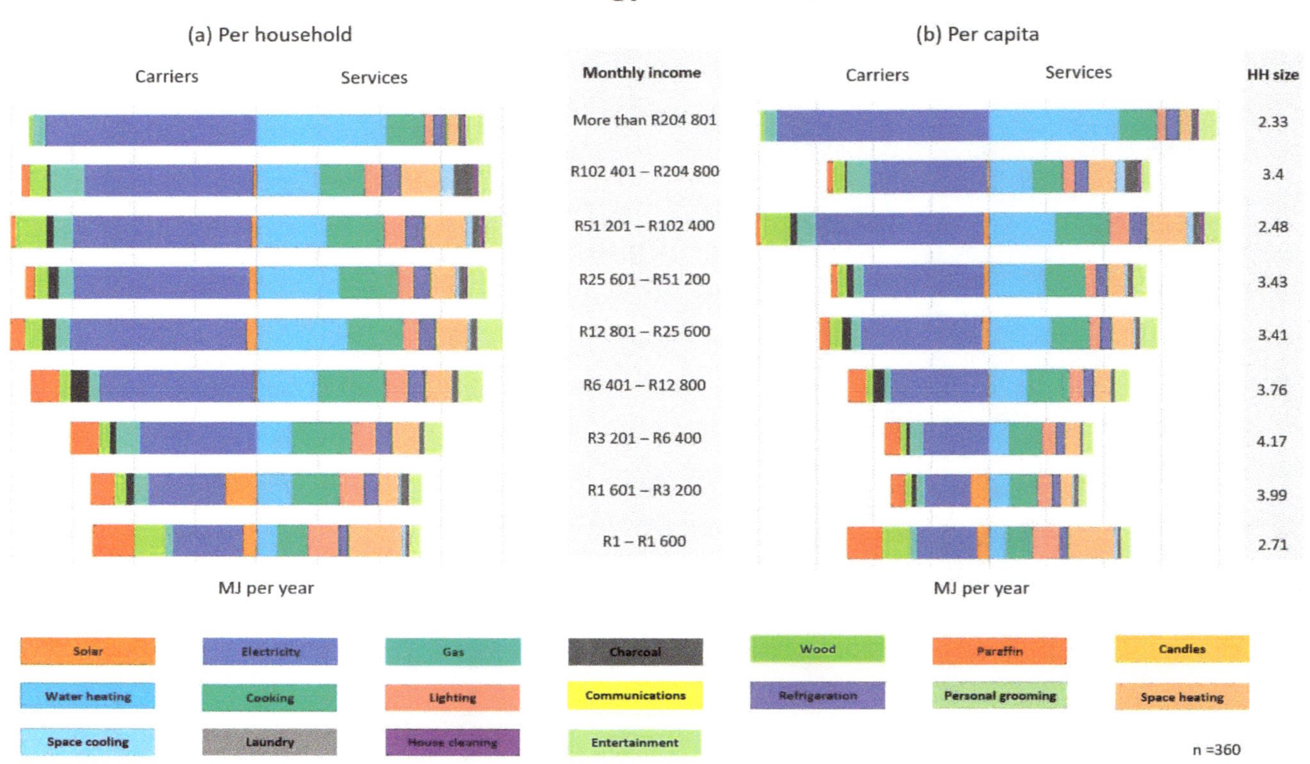

Figure 6. Household energy consumption categorized into carriers and services.

Figure 6 provided a basic understanding of how the consumption of carriers and services changes with monthly income, indicating that the method applied can provide this overview for a city, especially where a huge discrepancy exists in income levels. The monthly income levels display the income brackets that were presented to the respondents when answering the survey. However, it was also important to also understand more closely the relationship between carriers, services, and drivers for each income group in order to make more specific interventions for sustainability and to create the second pivotal conceptualization, namely diagrams that connect the carriers, services, and drivers for various income groups. These results and intervention points are discussed in detail in Strydom et al. (58). This section discusses how these results were reached.

In order to understand the inflows and through-flows of energy for various income groups, the first step was to create pie charts for each income group depicting the average carriers and services for that income group. Figure 7 depicts these pie charts for a low-income household in Cape Town.

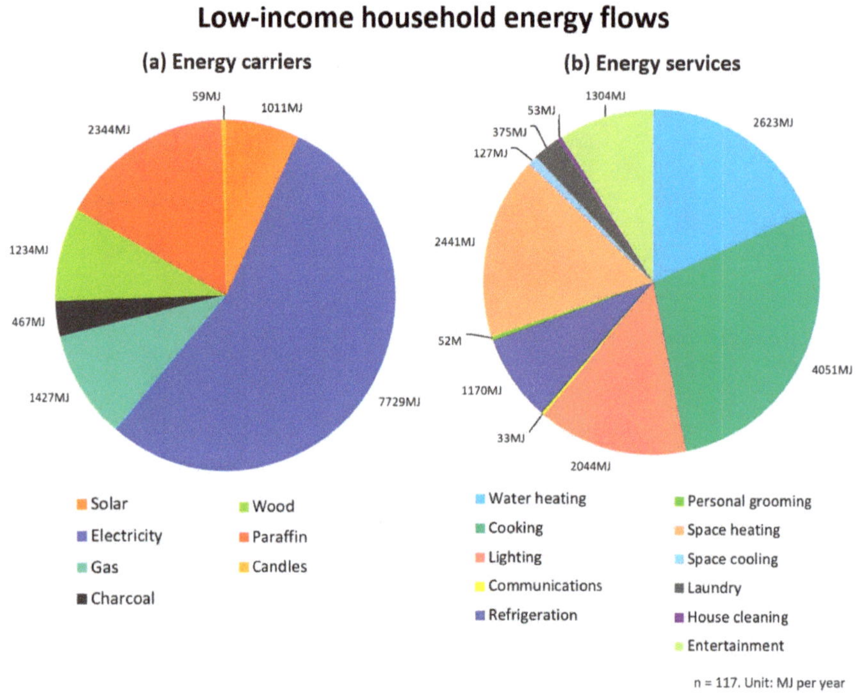

Figure 7. Average carriers and services for low-income households in Cape Town Source: Strydom et al. [58].

While Figure 7 provides insight into the size of various carriers and services, and could do so for other cities where this method is applied, it does not yet clarify the relationship between carriers and services. In order to understand carriers and services in terms of inflows and through-flows, Sankey diagrams were useful. Figure 8 is a Sankey diagram connecting carriers and services for a low-income household in Cape Town.

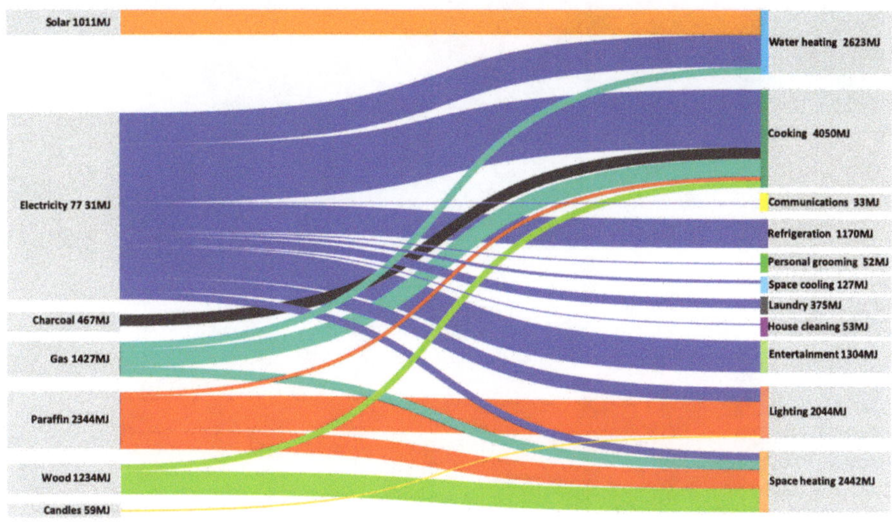

Figure 8. Average energy flows for low-income households in Cape Town. Source: Strydom et al. [58].

Figure 8 makes it possible to understand the relationship between carriers and services of the studied city, thereby making apparent intervention points for sustainability. In the case of Cape Town, it becomes clear that paraffin contributes greatly to lighting in low-income households. However, as paraffin is a very inefficient carrier, one requires much more paraffin to light a house than one would should electricity be the main lighting source. This means that replacing paraffin with electric light bulbs in low-income households will provide the same amount of lighting, if not more, while using less energy. Further insights and intervention points for Cape Town are discussed in full in Strydom et al. [58]. This diagram illustrates the potential insights into the household energy consumption of a specific income group that can be gained when applying the discussed method to a city.

In order to create the final conceptualization for energy consumption, it is important to connect energy carriers and services with energy drivers. Figure 9 provides this conceptualization for low-income households in Cape Town.

Figure 9 was created by totaling the energy services Sovacool [45] defined as satisfying subsistence needs and totaling the services defined as comfort, convenience, and cleanliness in another Sankey diagram. Creating such a diagram for each of the four income groups in Cape Town provided the opportunity for comparison and for understanding what drives energy consumption across income groups. The full results are discussed in Strydom et al. [58].

By viewing the two types of Sankey diagrams together, it is possible to arrive at a conceptualization of the relationship between carriers, services, and drivers for various income groups in cities across the world. Especially in the Global South, where low-income households consume a variety of energy carriers, it is possible to understand the importance of different energy carriers to fulfill different households' needs. Once such an understanding of household energy consumption is reached, it is possible to identify different intervention points for different income groups based on these energy flows. It is also possible to understand the core driver for accessing energy. In low-income households in Cape Town, it is clear that residents simply aim to satisfy subsistence needs, but that certain services associated with comfort, convenience, and cleanliness are important to them. By comparing these findings across income groups in other cities, it is possible to know what drives energy consumption and consequently which energy services a city's residents view as essential. With this knowledge, locally applicable intervention points for sustainability can be identified.

Figure 9. Energy drivers for low-income households. Source: Strydom et al. [58].

6. Conclusions

This paper reviewed the literature on urban metabolism, urban energy metabolism, and household energy metabolism to identify a conceptual framework with which to examine the inflows and through-flows of direct energy at the household level. The framework was further expanded to include drivers of energy consumption as they correlate with the services accessed in order to understand the level of access experienced in the household. It further examined the relationship between energy carriers and energy services based on income in Cape Town. The results show that both pie charts and Sankey diagrams provide the means to identify intervention points for sustainability.

This can in the future be applied to other cities where data is scarce and where residents access a range of energy carriers, not just electricity, as current studies assessing household energy consumption focus predominantly on electricity use and take national or city-level data to speak for household level energy consumption. It is flexible in that the carriers and services can be adapted depending the city studied. Since this study was conducted in a city that has both formal and informal housing, it also shows that the method can be used for both settings and for comparing results across formal and informal settlements. By using this method, it is possible to gain insight into the energy inflows and through-flows of households across income groups and dwelling types, and therefore provides a means to collect bottom–up data at the city level for a high-resolution image of the current state of energy consumption in our cities.

Author Contributions: Writing—original draft, conceptualization A.S.; methodology and data curation, A.S. and P.K.C.; supervision, writing, review and editing, J.K.M. and P.K.C.

Funding: This research was funded by the Centre for Renewable and Sustainable Energy Studies (CRSES), Stellenbosch, South Africa; and the National Research Foundation, South Africa, grant number 99103.

Conflicts of Interest: The authors declare no conflict of interest.

References

1. Barragán-Escandón, A.; Terrados-Cepeda, J.; Zalamea-León, E. The role of renewable energy in the promotion of circular urban metabolism. *Sustainability* **2017**, *9*, 2341. [CrossRef]
2. Grubler, A.; Bai, X.; Buettner, T.; Dhakal, S.; Fisk, D.; Ichinose, T.; Keirstead, J.; Sammer, G.; Satterthwaite, D.; Schulz, N. *Urban Energy Systems*; Cambridge University Press: Cambridge, UK; New York, NY, USA, 2012; pp. 1307–1400.
3. Barrera, P.P.; Carreón, J.R.; de Boer, H.J. A multi-level framework for metabolism in urban energy systems from an ecological perspective. *Resour. Conserv. Recycl.* **2018**, *132*, 230–238. [CrossRef]
4. Abrahamse, W.; Steg, L.; Vlek, C.; Rothengatter, T. A review of intervention studies aimed at household energy conservation. *J. Environ. Psychol.* **2005**, *25*, 273–291. [CrossRef]
5. Abrahamse, W.; Steg, L.; Vlek, C.; Rothengatter, T. The effect of tailored information, goal setting, and tailored feedback on household energy use, energy-related behaviors, and behavioral antecedents. *J. Environ. Psychol.* **2007**, *27*, 265–276. [CrossRef]
6. Banfi, S.; Farsi, M.; Filippini, M.; Jakob, M. Willingness to pay for energy-saving measures in residential buildings. *Energy Econ.* **2008**, *30*, 503–516. [CrossRef]
7. Benders, R.M.J.; Kok, R.; Moll, H.C.; Wiersma, G.; Noorman, K.J. New approaches for household energy conservation-In search of personal household energy budgets and energy reduction options. *Energy Policy* **2006**, *34*, 3612–3622. [CrossRef]
8. Kennedy, C.; Cuddihy, J.; Engel-yan, J. The Changing Metabolism of Cities. *J. Ind. Ecol.* **2007**, *11*, 43–59. [CrossRef]
9. McCalley, L.T.; Midden, C.J.H. Energy conservation through product-integrated feedback: The roles of goal-setting and social orientation. *J. Econ. Psychol.* **2002**, *23*, 589–603. [CrossRef]
10. Howells, M.I.; Alfstad, T.; Victor, D.G.; Goldstein, G.; Remme, U. A model of household energy services in a low-income rural African village. *Energy Policy* **2005**, *33*, 1833–1851. [CrossRef]
11. Musango, J.K.; Currie, P.; Robinson, B. *Urban Metabolism for Resource-Efficient Cities: From Theory to Implementation*; UN Environment: Paris, France, 2017; Available online: https://resourceefficientcities.org/wp-content/uploads/2017/09/Urban-Metabolism-for-Resource-Efficient-Cities.pdf (accessed on 22 August 2019).
12. Currie, P.K.; Musango, J.K.; May, N.D. Urban metabolism: A review with reference to Cape Town. *Cities* **2017**, *70*, 91–110. [CrossRef]
13. Thomson, G.; Newman, P. Urban fabrics and urban metabolism—From sustainable to regenerative cities. *Resour. Conserv. Recycl.* **2018**, *132*, 218–229. [CrossRef]
14. Currie, P.K.; Musango, J.K. African Urbanization: Assimilating Urban Metabolism into Sustainability Discourse and Practice. *J. Ind. Ecol.* **2016**, *21*, 1262–1276. [CrossRef]
15. Gouveia, J.P.; Seixas, J. Unraveling electricity consumption profiles in households through clusters: Combining smart meters and door-to-door surveys. *Energy Build.* **2016**, *116*, 666–676. [CrossRef]
16. Zhang, Y.; Yang, Z.; Yu, X. Urban Metabolism: A Review of Current Knowledge and Directions for Future Study. *Environ. Sci. Technol.* **2015**, *49*, 11247–11263. [CrossRef] [PubMed]
17. Salvini, P.; Miller, E.J. ILUTE: An Operational Prototype of a Comprehensive Microsimulation Model of Urban Systems. *Netw. Spat. Econ.* **2005**, *5*, 217–234. [CrossRef]
18. Mostafavi, N.; Farzinmoghadam, M.; Hoque, S. A framework for integrated urban metabolism analysis tool (IUMAT). *Build. Environ.* **2014**, *82*, 702–712. [CrossRef]
19. Patterson, Z.; Bierlaire, M. Development of Prototype Urbansim Models. *Environ. Plan. B Plan. Des.* **2010**, *37*, 344–366. [CrossRef]
20. Carreón, J.R.; Worrell, E. Urban energy systems within the transition to sustainable development. A research agenda for urban metabolism. *Resour. Conserv. Recycl.* **2018**, *132*, 258–266. [CrossRef]
21. Voskamp, I.M.; Spiller, M.; Stremke, S.; Bregt, A.K.; Vreugdenhil, C.; Rijnaarts, H.H.M. Space-time information analysis for resource-conscious urban planning and design: A stakeholder based identification of urban metabolism data gaps. *Resour. Conserv. Recycl.* **2018**, *128*, 516–525. [CrossRef]
22. Li, H.; Kwan, M.-P. Advancing analytical methods for urban metabolism studies. *Resour. Conserv. Recycl.* **2018**, *132*, 239–245. [CrossRef]
23. Lederer, J.; Kral, U. Theodor Weyl: A pioneer of urban metabolism studies. *J. Ind. Ecol.* **2015**, *19*, 695–702. [CrossRef]

24. Kennedy, C.; Pincetl, S.; Bunje, P. The study of urban metabolism and its applications to urban planning and design. *Environ. Pollut.* **2011**, *159*, 1965–1973. [CrossRef] [PubMed]

25. Currie, P.; Lay-Sleeper, E.; Fernández, J.E.; Kim, J.; Musango, J.K. Towards Urban Resource Flow Estimates in Data Scarce Environments: The Case of African Cities. *J. Environ. Prot.* **2015**, *6*, 1066–1083. [CrossRef]

26. Zhang, Y.; Li, S.; Fath, B.D.; Yang, Z.; Yang, N. Analysis of an urban energy metabolic system: Comparison of simple and complex model results. *Ecol. Model.* **2011**, *223*, 14–19. [CrossRef]

27. Chen, S.; Chen, B. Modelling carbon-energy metabolism of cities: A systems approach. *Energy Procedia* **2016**, *88*, 31–37. [CrossRef]

28. Zhang, Y.; Yang, Z.; Fath, B.D.; Li, S. Ecological network analysis of an urban energy metabolic system: Model development, and a case study of four Chinese cities. *Ecol. Model.* **2010**, *221*, 1865–1879. [CrossRef]

29. Bristow, D.N.; Kennedy, C.A. Urban metabolism and the energy stored in cities: Implications for resilience. *J. Ind. Ecol.* **2013**, *17*, 656–667. [CrossRef]

30. Fell, M. Energy services: A conceptual review. *Energy Res. Soc. Sci.* **2017**, *27*, 129–140. [CrossRef]

31. Zhang, Y.; Zheng, H.; Fath, B.D. Analysis of the energy metabolism of urban socioeconomic sectors and the associated carbon footprints: Model development and a case study for Beijing. *Energy Policy* **2014**, *73*, 540–551. [CrossRef]

32. Weisz, H.; Steinberger, J.K. Reducing energy and material flows in cities. *Environ. Sustain.* **2010**, *2*, 185–192. [CrossRef]

33. Leduc, W.R.W.A.; Van Kann, F.M.G. Spatial planning based on urban energy harvesting toward productive urban regions. *J. Clean. Prod.* **2013**, *39*, 180–190. [CrossRef]

34. Brunner, P.H. Reshaping urban metabolism. *J. Ind. Ecol.* **2007**, *11*, 11–13. [CrossRef]

35. Giampietro, M.; Mayumi, K. Multiple-scale integrated assessments of societal metabolism: Integrating biophysical and economic representations across scales. *Popul. Environ.* **2000**, *22*, 155–210. [CrossRef]

36. Moll, H.C.; Noorman, K.J.; Kok, R.; Engström, R.; Throne-Holst, H.; Clark, C. Pursuing More Sustainable Consumption by Analyzing Household Metabolism in European Countries and Cities. *J. Ind. Ecol.* **2005**, *9*, 259–275. [CrossRef]

37. Biesiot, W.; Noorman, K.J. Energy requirements of household consumption: A case study of The Netherlands. *Ecol. Econ.* **1999**, *28*, 367–383. [CrossRef]

38. Di Donato, M.; Lomas, P.L.; Carpintero, Ó. Metabolism and environmental impacts of household consumption: A review on the assessment, methodology, and drivers. *J. Ind. Ecol.* **2015**, *19*, 904–916. [CrossRef]

39. Yang, D.; Gao, L.; Xiao, L.; Wang, R. Cross-boundary environmental effects of urban household metabolism based on an urban spatial conceptual framework: A comparative case of Xiamen. *J. Clean. Prod.* **2012**, *27*, 1–10. [CrossRef]

40. Frostell, B.M.; Sinha, R.; Assefa, G.; Olsson, L.E. Modeling both direct and indirect environmental load of purchase decisions: A web-based tool addressing household metabolism. *Environ. Model. Softw.* **2015**, *71*, 138–147. [CrossRef]

41. Elkhorchani, H.; Grayaa, K. Novel home energy management system using wireless communication technologies for carbon emission reduction within a smart grid. *J. Clean. Prod.* **2016**, *135*, 950–962. [CrossRef]

42. Shakeri, M.; Shayestegan, M.; Abunima, H.; Reza, S.M.S.; Akhtaruzzaman, M.; Alamoud, A.R.M.; Sopian, K.; Amin, N. An intelligent system architecture in home energy management systems (HEMS) for efficient demand response in smart grid. *Energy Build.* **2017**, *138*, 154–164. [CrossRef]

43. Zhou, B.; Li, W.; Chan, K.W.; Cao, Y.; Kuang, Y.; Liu, X.; Wang, X. Smart home energy management systems: Concept, configurations, and scheduling strategies. *Renew. Sustain. Energy Rev.* **2016**, *61*, 30–40. [CrossRef]

44. De Almeida, A.; Fonseca, P.; Schlomann, B.; Feilberg, N. Characterization of the household electricity consumption in the EU, potential energy savings and specific policy recommendations. *Energy Build.* **2011**, *43*, 1884–1894. [CrossRef]

45. Sovacool, B.K. Conceptualizing urban household energy use: Climbing the "Energy Services Ladder". *Energy Policy* **2011**, *39*, 1659–1668. [CrossRef]

46. Kwak, S.Y.; Yoo, S.H.; Kwak, S.J. Valuing energy-saving measures in residential buildings: A choice experiment study. *Energy Policy* **2010**, *38*, 673–677. [CrossRef]

47. Salo, M.; Nissinen, A.; Lilja, R.; Olkanen, E.; O'Neill, M.; Uotinen, M. Tailored advice and services to enhance sustainable household consumption in Finland. *J. Clean. Prod.* **2016**, *121*, 200–207. [CrossRef]

48. Kowsari, R.; Zerriffi, H. Three dimensional energy profile: A conceptual framework for assessing household energy use. *Energy Policy* **2011**, *39*, 7505–7517. [CrossRef]
49. Bouzarovski, S.; Petrova, S. A global perspective on domestic energy deprivation: Overcoming the energy poverty-fuel poverty binary. *Energy Res. Soc. Sci.* **2015**, *10*, 31–40. [CrossRef]
50. Gillard, R.; Snell, C.; Bevan, M. Advancing an energy justice perspective of fuel poverty: Household vulnerability and domestic retrofit policy in the United Kingdom. *Energy Res. Soc. Sci.* **2017**, *29*, 53–61. [CrossRef]
51. Middlemiss, L.; Gillard, R. Fuel poverty from the bottom-up: Characterising household energy vulnerability through the lived experience of the fuel poor. *Energy Res. Soc. Sci.* **2015**, *6*, 146–154. [CrossRef]
52. Camara, N.F.; Xu, D.; Binyet, E. Understanding household energy use, decision making and behaviour in Guinea-Conakry by applying behavioural economics. *Renew. Sustain. Energy Rev.* **2017**, *79*, 1380–1391. [CrossRef]
53. Budya, H.; Yasir Arofat, M. Providing cleaner energy access in Indonesia through the megaproject of kerosene conversion to LPG. *Energy Policy* **2011**, *39*, 7575–7586. [CrossRef]
54. Foell, W.; Pachauri, S.; Spreng, D.; Zerriffi, H. Household cooking fuels and technologies in developing economies. *Energy Policy* **2011**, *39*, 7487–7496. [CrossRef]
55. Maes, W.H.; Verbist, B. Increasing the sustainability of household cooking in developing countries: Policy implications. *Renew. Sustain. Energy Rev.* **2012**, *16*, 4204–4221. [CrossRef]
56. Parikh, J. Hardships and health impacts on women due to traditional cooking fuels: A case study of Himachal Pradesh, India. *Energy Policy* **2011**, *39*, 7587–7594. [CrossRef]
57. Williams, K.N.; Northcross, A.L.; Graham, J.P. Health impacts of household energy use: Indicators of exposure to air pollution and other risks. *Perspectives* **2015**, *93*, 507–508. [CrossRef] [PubMed]
58. Strydom, A.; Musango, J.K.; Currie, P.K. Connecting energy services, carriers and flows: rethinking household energy metabolism in Cape Town, South Africa. *Energy Res. Soc. Sci.* Forthcoming. [CrossRef]

 *energies*

Article

Community Resilience-Oriented Optimal Micro-Grid Capacity Expansion Planning: The Case of Totarabank Eco-Village, New Zealand

Soheil Mohseni [1,*] , Alan C. Brent [1,2] and Daniel Burmester [1]

[1] Sustainable Energy Systems, School of Engineering and Computer Science, Faculty of Engineering, Victoria University of Wellington, PO Box 600, Wellington 6140, New Zealand; alan.brent@vuw.ac.nz (A.C.B.); daniel.burmester@vuw.ac.nz (D.B.)

[2] Department of Industrial Engineering and the Centre for Renewable and Sustainable Energy Studies, Stellenbosch University, Stellenbosch 7600, South Africa

*   Correspondence: soheil.mohseni@ecs.vuw.ac.nz

Received: 1 June 2020; Accepted: 18 July 2020; Published: 2 August 2020

Abstract: In the grid-tied micro-grid context, energy resilience can be defined as the time period that a local energy system can supply the critical loads during an unplanned upstream grid outage. While the role of renewable-based micro-grids in enhancing communities' energy resilience is well-appreciated, the academic literature on the techno-economic optimisation of community-scale micro-grids lacks a quantitative decision support analysis concerning the inclusion of a minimum resilience constraint in the optimisation process. Utilising a specifically-developed, time-based resilience capacity characterisation method to quantify the sustainability of micro-grids in the face of different levels of extended grid power outages, this paper facilitates stakeholder decision-making on the trade-off between the whole-life cost of a community micro-grid system and its degree of resilience. Furthermore, this paper focuses on energy infrastructure expansion planning, aiming to analyse the importance of micro-grid reinforcement to meet new sources of electricity demand—particularly, transport electrification—in addition to the business-as-usual demand growth. Using quantitative case study evidence from the Totarabank Subdivision in New Zealand, the paper concludes that at the current feed-in-tariff rate (NZ$0.08/kWh), the life cycle profitability of resilience-oriented community micro-grid capacity reinforcement is guaranteed within a New Zealand context, though constrained by capital requirements.

Keywords: renewable energy systems; microgrids; optimal expansion planning; energy resilience; resilient energy systems; critical loads; electric vehicles; techno-economic analysis; HOMER Pro; New Zealand

---

1. Introduction

The past two decades have witnessed a remarkable evolution of micro-grids (MGs) from a nascent concept to a pivotal player in the transition to 100% renewable energy [1–3]. The power industry has accordingly seen an ever-increasing penetration of distributed energy resources into utility grids. The key drivers behind the stakeholders' willingness towards additional, often non-trivial, capital investments for resilience capacity are [4–6]: (1) the constantly falling costs of renewable energy technologies, which are transforming the economics of green energy, most notably solar photovoltaic (PV) panels and battery energy storage systems (BESSs), (2) the growing awareness about the inadequacy of current approaches to energy resilience at community-scale, specifically the use of diesel generators to provide backup power during utility grid outages, and (3) the growth of engaged prosumers, who place great value on their energy self-sufficiency for sustainable living.

In the context of on-grid MGs, energy resilience is generally defined as "the amount of time that an MG can sustain critical loads during a grid outage", and often serves as a synonym for outage survivability [7]. A recent, growing strand of the literature has documented and emphasised the increasingly important role of planning for resilience within the context of grid-tied MGs, which, in turn, improves the resilience of a nation's entire power grid. For instance, Eskandarpour et al. [8] have formulated a mixed-integer linear programming problem for the optimal sizing and siting of MGs within power systems by considering the cost of unserved energy during utility grid outages as the objective function, while adhering to a limited budget for resilience improvements. In another instance, Barnes et al. [9] have revealed the potentially significant benefits of including energy resilience constraints in the long-term investment planning problem of networked MGs in terms of the prolonged outage survivability and overall cost-efficiency. Furthermore, Anderson et al. [7] have shown the economic and resilience benefits delivered by renewable energy technologies in a hybrid MG. More specifically, they have demonstrated that adding a 845-kW PV system together with a 172-kWh BESS to an existing 305 kW of backup diesel generators extends the time period the MG can sustain crucial loads by 1.8 days, while generating savings of US$104,000 in energy costs over the reinforced MG life-cycle. Table 1 provides an overview of recent studies centred on the optimal capacity expansion planning of grid-tied renewable energy systems. Moreover, Table 2 presents a summary of the previous studies that focused on the long-term, integrated resource and resilience planning in the context of renewable and sustainable energy systems (in addition to the studies reviewed above). The reader is referred to [10] for a more detailed review of the methods and research trends in the MG resilience improvement literature.

As Tables 1 and 2 indicate, no previous study, as far as can be ascertained, has introduced a systematic method to quantify the cost of insuring grid-tied renewable energy systems against sustained grid outages in the optimal equipment capacity expansion planning processes—to meet some desired minimum level of resilience. Accordingly, this paper bridges the gap between these two streams of literature by proposing a modelling framework to estimate the additional costs incurred to procure energy resilience in the optimal capacity expansion planning processes of grid-connected community MGs. The paper additionally presents a sensitivity analysis to highlight the impact of varying minimum degrees of energy resilience required by stakeholders on the optimal life-cycle costs associated with the capacity reinforcement of grid-connected MGs.

The remainder of this paper is organised as follows. Section 2 describes the overall structure of the test-case, the fundamental assumptions underlying the study, and data requirements for the case study. Section 3 provides the modelling approach, while Section 4 presents the results and examines the robustness of optimal MG capacity configurations. Finally, conclusions are made and potential areas for future work are discussed in Section 5.

Table 1. Summary of the recent previous work dealing with the cost-optimal capacity expansion planning of renewable energy systems.

Ref.	Technologies Considered in the Candidate Pool	Optimisation Approach	Key Contribution(s)	Key Insight(s) from Case Study Analyses
[11]	Solar photovoltaic (PV), wind turbine, fuel cell, diesel generator, heat storage, and battery	Mixed-integer linear programming solved using the General Algebraic Modeling System (GAMS)	• Using interval linear programming to characterise the uncertainty associated with long-term weather projections. • Taking a multi-period planning approach to improve the precision of decision-making in the allocation of renewable energy resources.	• The technology diversity degree decreases, as the acceptable level of robustness against weather forecast uncertainties increases. • At present, fuel cells are not cost-competitive with mature battery technologies.
[12]	Solar PV, wind turbine, diesel generator, and battery	Scenario-based stochastic optimisation	• Considering multiple conflicting objectives, namely the minimisation of life-cycle costs, greenhouse gas emissions, and the non-renewable fraction of power generation capacity.	• The total net present cost of capacity additions would have been underestimated by about 23% if the model-inherent uncertainties (forecasts of load demand and climatic variables) were not modelled.
[13]	Wind turbine, bi-directional electric vehicle charging infrastructure	Semi-dynamic, semi-static programming	• Formulating the utility generation expansion planning problem to determine the optimal additional capacity of wind generation and the optimal penetration of electric vehicles that maximise the utility's profits in the long run.	• Integration of vehicle-to-grid technology can play a significant role in improving the economic viability of renewable energy capacity expansion decisions.

Table 1. *Cont.*

Ref.	Technologies Considered in the Candidate Pool	Optimisation Approach	Key Contribution(s)	Key Insight(s) from Case Study Analyses
[14]	Solar PV, wind turbine, diesel generator, battery, and utility line	Mixed-integer nonlinear programming solved using meta-heuristic optimisation algorithms	• Simultaneous optimisation of long-term investment costs and short-term operating costs.	• Co-optimisation of investment planning and energy scheduling problems reduces the total life-cycle costs by about 10%. • Increasing the capacity of the line connecting the micro-grid (MG) to the upstream grid results in lower costs of handling the uncertainties in load and weather forecasts.
[15]	Solar PV and wind turbine	Multi-agent systems	• Bi-layer, agent-based simulation of the capacity expansion planning problem to aid MG operators in decision-making among a range of candidate investment plans to maximise their profits. The inner (operational) layer simulates the renewable energy market behaviour, the outputs of which are fed into the outer (planning) layer, which evaluates the profitability of each investment scenario.	• Characterising the competitive behaviour of MG operators within an existing utility's territory makes the numeric simulation results more accurate and representative of real-world practice.

Table 2. Summary of the previous work focused on the integration of resilience constraints into the long-term investment planning processes of renewable energy systems.

Ref.	Technologies/Resources/Systems Considered in the Candidate Pool	Optimisation Approach	Key Contribution(s)	Key Insight(s) from Case Study Analyses
[16]	Unspecified distributed generation	Mixed-integer linear programming solved using GAMS	• Treating the uncertainty associated with the occurrence of extreme weather events using information gap decision theory. • Proposing a resilience planning model for active distribution grids to produce optimal trade-offs between the added capacity of backup distributed generation and hardened utility lines within certain budget constraints.	• Information gap decision theory is an effective approach for supporting decision-making in the planning of distribution electricity networks for resilience. • The best-compromise solution between additional distribution generation allocation and distribution line hardening highly relies on the acceptable degree of conservativeness for the resilience of the system.
[17]	Hydro, wind, solar, geothermal, nuclear, coal, natural gas, and oil	Multi-objective optimisation of the national-level, energy infrastructure capacity expansion problem	• Assessing the resilience of the U.S. national energy and transportation systems to a set of extreme events. • Considering cost, sustainability, and resilience as three independent objectives. • Proposing a resilience measure to quantify the ability of the energy and transportation systems to recover from the consequences of a large-catastrophic event.	• Investment and operating costs, sustainability degree, and resilience level are competing objectives, which cannot be improved at the same time, when jointly planning national energy and transportation systems.

Table 2. *Cont.*

Ref.	Technologies/Resources/Systems Considered in the Candidate Pool	Optimisation Approach	Key Contribution(s)	Key Insight(s) from Case Study Analyses
[18]	Electric power system and natural gas system	Two-stage robust optimisation	• Proposing an integrated expansion planning framework for electricity and natural gas systems, which aims at improving power grid resilience against extreme events.	• Joint investment planning of electricity and natural gas networks provides the opportunity to significantly reduce the cost of resilience as compared to the case where resilience requirements are met alone through additional power grid resources.
[19]	Solar PV, battery, and combined heat and power	HOMER software	• Characterising the desired resilience preferences of a site based on the battery bank's duration of autonomy. Specifically, battery banks were sized to serve the peak load demand for 4 h in cases of a grid outage.	• The expected costs for the additional battery bank to meet the resilience requirements scale with the size of the project (that can be measured by the mean load demand). • Achieving resilience requirements using additional battery arrays incurs sizable capital, replacement, and operation and maintenance costs.
[20]	Unspecified distributed generation technologies	Two-stage robust optimisation	• Proposing an optimal distributed generation placement model that minimises the total load shedding in extreme events, whilst additionally characterising the uncertainty coupled with the time of occurrence and duration of natural disasters.	• Increasing the penetration of distributed generation resources within distribution networks significantly improves the cost-efficiency of procuring resilience provisions.

Table 2. *Cont.*

Ref.	Technologies/Resources/Systems Considered in the Candidate Pool	Optimisation Approach	Key Contribution(s)	Key Insight(s) from Case Study Analyses
[21]	Solar PV, wind turbine, diesel generator, and battery	Two-stage robust optimisation	• Proposing a grid-tied MG investment planning model to optimally site and size candidate sets of distributed energy resources and utility lines, whilst adhering to a pre-specified degree of resilience to N-k contingencies. [1]	• Increasing the minimum required resilience level increases the life-cycle cost of grid-connected MGs. • The computational time increases exponentially with the maximum number of simultaneous contingencies considered.
[22]	Battery and diesel generator	Mixed-integer linear programming solved using GAMS	• Developing a framework to optimise the type and size of backup power equipment for critical services in islanded operation modes of grid-tied MGs. • Measuring the end-consumer preferences for resilience to grid outages using the average critical customer interruption index. • Quantification of the stochasticity associated with extreme weather events in terms of occurrence time and duration.	• Capturing the stochasticity of extreme weather events could reduce the cost of MG capacity reinforcement—through dedicated additional backup power capacity—to meet the resilience requirements by up to about 12%, as compared with deterministic models.

[1] N-k contingency criterion guarantees that N critical components within a power network can continue normal operation in cases any k components simultaneously suffer a failure.

2. Test-Case System: Totarabank, New Zealand

The Totarabank Subdivision consists of eight freehold residential lots (ranging in size from 1200 m^2 to 2100 m^2) and a communal building (on the ninth, common lot) with a large area of around 6 ha held in common ownership [23]. Located in central Wairarapa, New Zealand (GPS coordinates: 41°1′4″ S 175°40′0″ E), the subdivision adheres to sustainability design criteria to suggest ways for resilience integration. Figure 1 shows the geographical location and description of the case study subdivision Totarabank.

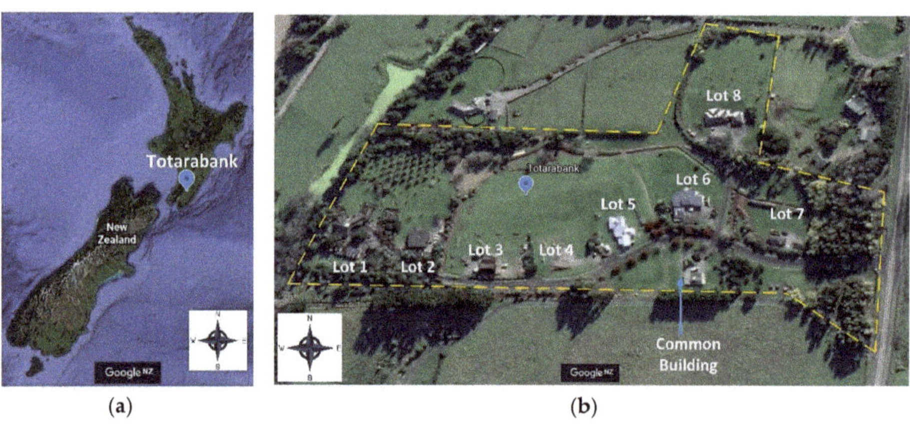

(a)   (b)

Figure 1. Satellite photographs of the case study area: (**a**) location of the subdivision on a New Zealand map; (**b**) layout of the subdivision showing the lots (image courtesy of Google Earth).

As of May 2020, Totarabank has 14 inhabitants, and an existing installed grid-integrated power generation capacity of 11.4 kW, 100% from solar PV modules, which have an average remaining life span of 19.25 years. Accordingly, the load power demand is met through electricity importing from the main grid at night when the onsite PV system is not delivering, while any surplus electricity is sold back to the grid during the day. However, the primary issue of the current power dispatching strategy is its failure to serve the crucial loads at night if the upstream network fails or requires maintenance. This, together with the anticipated growth in demand (led by the electrification of transport), reveals the necessity for the reinforcement of the current energy system—to improve the site's energy resilience, self-sufficiency, and security of supply.

2.1. Candidate Technologies

The considered site is richly endowed with renewable energy resources, particularly solar and wind. The candidate technologies included in the model for optimisation are (1) PV modules, (2) wind turbines (WTs), and (3) a BESS. Furthermore, an unreported preliminary techno-economic study suggested that a BESS is the most cost-effective technology for onsite energy storage among feasible options for Totarabank, namely hydrogen storage system (electrolyser, hydrogen reservoir, and fuel cell), batteries, flywheels, and super-capacitors. More specifically, a Li-ion battery system was chosen, as it offers the best combination of energy density, low self-discharge, and manufacturing cost, compared with other mature battery technologies [24]. Accordingly, the selected components were assembled in an AC-coupled configuration to form the grid-connected MG shown in Figure 2.

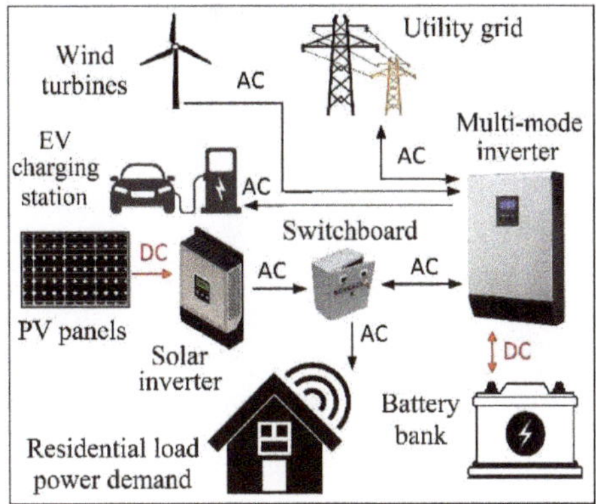

Figure 2. Schematic of the conceptualised grid-tied, AC-coupled MG system.

2.1.1. PV Modules

The power output from each PV module in time t is obtained from Equation (1) as a function of solar irradiance and the module's temperature, which is estimated from Equation (2) [25].

$$P_{PV}^t = P_{PV,r} \times \eta_{PV} \times f_{PV} \times \left(\frac{G_T^t}{G_{T,STC}} \right) \times \left(1 - \left(\frac{k_p}{100} \times \left(T_{PV}^t - T_{STC} \right) \right) \right), \tag{1}$$

$$T_{PV}^t = T_a^t + G_T^t \times \left(\frac{NOCT - 20}{0.8} \right), \tag{2}$$

where $P_{PV,r}$, η_{PV}, f_{PV}, k_p, and $NOCT$ denote the module's rated power, efficiency, derating factor, temperature coefficient, and nominal operating cell temperature, respectively; G_T^t is the global horizontal irradiance in time t; $G_{T,STC}$ and T_{STC} represent the solar irradiance and temperature at the standard test conditions, respectively; and T_a^t is the ambient temperature.

2.1.2. Wind Turbines

The power output from each WT is obtained from the manufacturer-provided power curve, shown in Figure 3 [26]. Additionally, the wind speed data is hub height normalised using the power-law, which is expressed in Equation (3) [27].

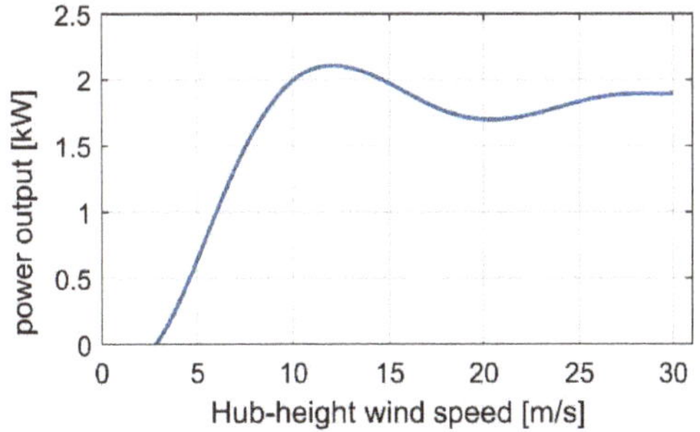

Figure 3. Power curve of the selected wind turbine. Data source: [26].

$$V_{hub}^t = V_{ref}^t \left(\frac{h_{hub}}{h_{ref}} \right)^{\gamma}, \tag{3}$$

where V_{ref} represents wind speed measured at the height of h_{ref} in time t, and γ is the wind shear exponent.

2.1.3. Battery Arrays

The energy content of the battery bank, comprising of battery packs connected in parallel, in time t can be calculated from Equation (4) [28].

$$E_B^t = E_B^{t-\Delta t} + \left(P_B^t \times \Delta t\right), \tag{4}$$

where Δt denotes the operating time increment and P_B^t is the charging/discharging power of the battery bank in time t that can be calculated based on the system-wide supply-demand balance constraint from Equation (5) [29].

$$P_B^t = P_{PV}^t + P_{WT}^t \mp P_g^t - P_L^t, \tag{5}$$

where P_{WT}^t is the power output from WTs in time t, P_g^t is the exchanged power with the utility grid, a positive (negative) value of which represents an imported (exported) power, and P_L^t is the total residential load power demand. A positive value of P_B^t represents the charging state, whereas negative values signify the discharging state.

Moreover, the energy stored in each battery pack is constrained to lie within the feasible range of values, $[E_{B,min}, E_{B,max}]$.

2.1.4. Hybrid Inverter

A grid-interactive inverter (also called multi-mode battery inverter/charger, or bidirectional dual-mode hybrid inverter) is used in an AC-coupled configuration, which is capable of managing inputs from multiple sources in both on- and off-grid operating modes. The hybrid inverter is modelled by its overall efficiency.

2.1.5. Utility Grid

The cost of importing/exporting electricity from/to the utility grid in time t is obtained from Equation (6) [30].

$$cost_g^t = \begin{cases} \pi_{ex}^t \times P_g^t \times \Delta t & if\ P_g^t > 0, \\ FiT \times P_g^t \times \Delta t & if\ P_g^t < 0, \end{cases} \tag{6}$$

where π_{ex}^t is the wholesale power price in time t and FiT represents a fixed feed-in-tariff.

2.1.6. Electric Vehicle Charging Station

AC level 2 charging, which is the most common home charging solution, is considered in this project [31]. Accordingly, a level 2 electric vehicle supply equipment (EVSE) is employed, which is modelled by its efficiency. It is connected directly to the MG's hybrid inverter via a dedicated circuit and provides charging through a 240 V AC plug.

2.2. Key Assumptions

The following simplifying assumptions were made:

- The MG capacity expansion planning was carried out from a macro (centralised) perspective. Accordingly, this study does not focus on how to optimally assign the equipment capacity to each lot.
- The costs associated with the replacement and operation and maintenance (O&M) of the existing installed solar inverters were not reflected in the model. The reason lies in the fact that these assets are privately owned, while the new capacity additions were assumed to be shared by the community. That is, accounting for the replacement and O&M costs of the currently privately held

solar inverters will require more sophisticated market designs (such as peer-to-peer markets) for the intra-community electricity exchanges to establish a fair playing field—which are currently cleared under a pure flat-rate tariff structure. Note that the rationale behind making this assumption stems from the difference in the service life of PV panels and solar inverters.

- The energy stored in the stationary battery bank is not allowed to be used for electric vehicle (EV) charging purposes—allowed only for critical loads to address the resilience of the community and to improve the service life of the stationary battery bank.

- The case study site's mobility requirements were assumed to be 40 km/day/lot, while the EVs' average efficiency was assumed to be 0.12 kWh/km, considering a Nissan Leaf, which is New Zealand's most popular EV [32–34].

- A cooperative energy scheduling strategy (for example the one proposed in [35]), to be materialised in the implementation phase, was assumed to be able to coordinate the flexible charging of EVs, such that the daily periods of time the EV charging infrastructure sits unused is minimised.

- Vehicle-to-grid (V2G) services [36], as well as the effect of load growth due to the eco-village's population growth, were not taken into consideration. Rather, the system expansion is planned to meet the expectation of growing loads from the existing number of inhabitants.

- The product models were chosen, based on the authors' judgement of both efficiency and cost-effectiveness, from the options available in the Australia and New Zealand renewable energy markets, while costs are always cited in New Zealand currency.

2.3. Data

The techno-economic specifications of the selected components and the associated sources are summarised in Table 3.

Table 3. Techno-economic specifications of the components.

Specification	Component			
	PV Modules [1]	Wind Turbines [2]	Battery Arrays	Converter
Manufacturer part number	TSM-285 PD05, Trina Solar	X-2000L	RESU 3.3, LG Chem	SPMC240-AU, Selectronic
Rated capacity	285 W	2 kW	3.3 kWh	3 kW
Capital cost	$237/unit $832/kW	$3967/unit $1984/kW	$3645/unit $1105/kWh	$4600/unit $1533/kW
Replacement cost [3]	$237/unit	$3229/unit	$3645/unit	$4600/unit
O&M cost [3]	$0.7/unit/year	$26.4/unit/year	$7.3/unit/year	$3.9/unit/year
Useful life	25 years	20 years [4]	15 years	15 years
Efficiency	17.4%	N/A [5]	95% [6]	96%
Source	[37,38]	[26]	[39–41]	[42,43]

[1] Photovoltaic (PV) module costs include the cost associated with the solar inverter. [2] WT costs include the cost of the required guyed tower. [3] The replacement and O&M costs were adjusted in accordance with the capital-to-replacement and capital-to-O&M cost ratios used in [44–46]. [4] The approximate average operational service life of small and micro WTs is 120,000 h. However, since they do not operate at wind speeds below cut-in, their service life is typically considered as 20 years [47]. [5] N/A: Not applicable, since the WT's performance is modelled by its characteristic power-speed curve, in the same way as in [48]. [6] The battery bank's efficiency represents its round-trip efficiency.

Moreover, in view of the study objectives and simplifying assumptions outlined above, the costs associated with the EVSE were exogenously treated. More specifically, they were taken into account through cost premiums imposed on the optimised MG whole-life cost solutions. Specifically, the Delta AC Mini Plus EVSE, was chosen in this analysis to serve the level 2 charging. It has the following techno-economic specifications [49,50]: capital and replacement costs = $2564/unit, O&M cost = $17/unit/year, service life = 25 years, efficiency = 99%, maximum output power = 7.36 kW, and input rating = 230 V AC. Accordingly, the EV charging load was classified as deferrable, the monthly average value of which was assumed to be constant throughout the year, with an average value of 48 kWh/d, and a peak value of $(n \times 7.36)$ kW, in compliance with the maximum power output from n EV chargers.

Table 4 presents the data and sources for the parameter settings of the conceptualised MG system.

Table 4. Conceptualised MG system's scalars and sources.

Scalar	Value	Source	Scalar	Value	Source
f_{PV}	88%	[51]	h_{hub}	7 m	(this paper)
k_p	−0.41%/°C	[37]	h_{ref}	50 m	[52]
$NOCT$	44	[37]	FiT	\$0.08/kWh	[53]
$G_{T,STC}$	1 kW/m^2	[54]	Δt	1 h	[55]
T_{STC}	25 °C	[54]	$E_{B,max}^{\ 1}$	2.9 kWh	[40]
γ	0.15	[56]	$E_{B,min}^{\ 2}$	0.29 kWh	[40]

$^1 E_{B,max}$ denotes the maximum usable capacity of each module. ^{2} By definition, $E_{B,min} = (1 - DOD) \times E_{B,max}$, where DOD is the depth of discharge [%] [35].

Figure 4 shows the average wholesale power price for each hour across the five-year period of 2015 to 2019 [57], which were considered as forecasts of utility power price.

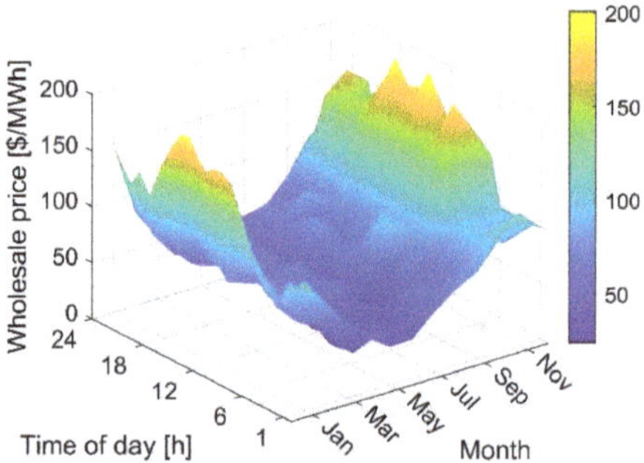

Figure 4. Forecasted profile for the wholesale electricity price at hourly resolution. The numerical values belonging to this figure are listed in Table S1 in Supplementary Material.

Figure 5 depicts the forecasted residential load power demand profile. The time-series data were synthesised from the GREEN Grid household electricity demand data [58] through downscaling the GREEN Grid study's sample of dwellings to the Totarabank's household population according to the proportionate scaling method described in [59].

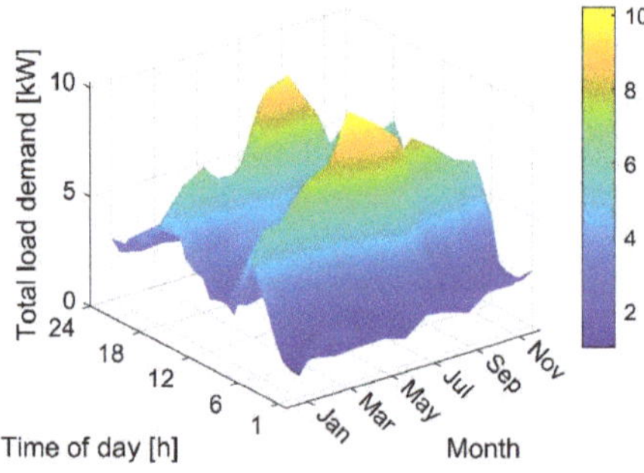

Figure 5. Forecasted profile for residential loads at hourly resolution. The numerical values belonging to this figure are listed in Table S2 in Supplementary Material.

Forecasts of wind and solar resources were determined using the NASA Surface Meteorology and Solar Energy database [52]. Accordingly, the monthly-averaged wind speed values over a 10-year historical period, and monthly-averaged insolation and air temperature values over a 22-year historical period were retrieved from the database. Monthly mean forecasts of wind speed and solar radiation are shown in Figure 6, while Figure 7 displays monthly mean temperature [52]. In general, an annual average wind speed value of 5 m/s, and an annual insolation value of 4 kWh/m^2/d are commonly considered acceptable thresholds for the commercial viability of wind and solar projects, while seasonal variations of resources can also affect the optimal resource combination of a PV/WT renewable energy system [60]. As can be seen from Figure 6, not only is wind energy a more reliable source of power than solar in the considered area, but it also shows a more constant trend than solar energy.

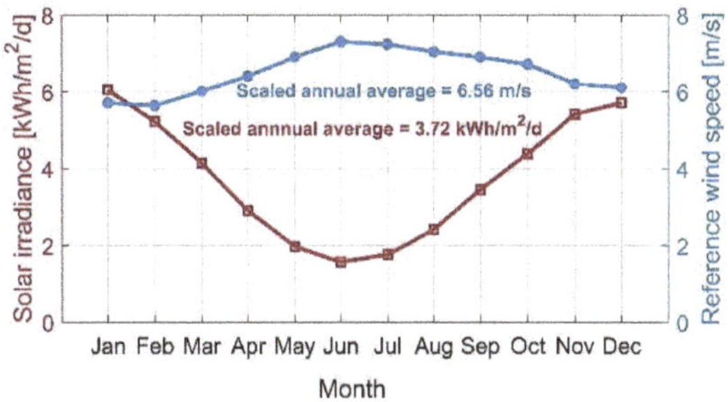

Figure 6. Forecasted monthly average insolation and wind speed data.

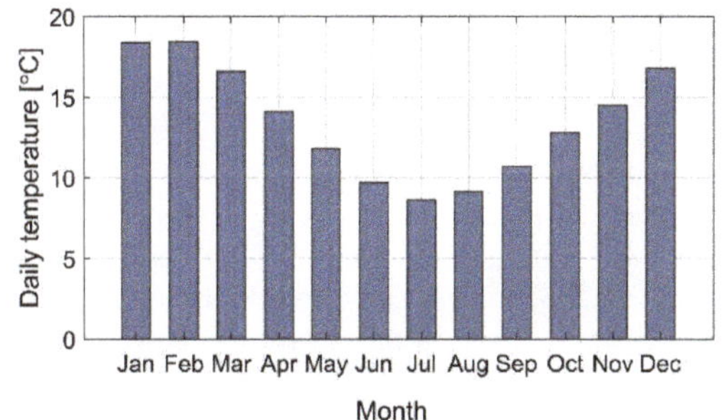

Figure 7. Forecasted monthly average ambient temperature data.

3. Methodology

This section presents the methodology developed for the resilience-constrained capacity expansion planning of MGs.

3.1. Modelling Approach

The U.S. National Renewable Energy Lab's Hybrid Optimization of Multiple Energy Resources (HOMER) Pro software (version 3.13.8) [61] was used to minimise the project's life-cycle cost subject to techno-economic design constraints for different levels of energy resilience. The software features two algorithms to search for the cost-optimal design configuration: (1) an original grid search algorithm that simulates all of the feasible resource combinations that meet the loads, and (2) a new proprietary derivative-free optimisation algorithm, called "HOMER Optimizer", which frees the user from the need to define the search space. Additionally, the conceptualised MG's life-cycle cost reflects all of the cost components associated with supplying energy to the considered site in present value, namely, the capital, replacement, and O&M costs of new energy infrastructure, as well as the net cost of energy exchange with the utility grid. At each time-step of the MG operation, an energy dispatch analysis is carried out, whereby the load demand is served by the least-cost combination of dispatchable and non-dispatchable distributed energy generation units, released energy from storage devices, traded energy with the upstream grid, and load shifting. Additionally, deferrable loads are placed in second priority, and supplied using the surplus power (to the needs of primary loads) from non-dispatchable micro-generation units—solar PV and WT technologies—ahead of charging the storage devices. The energy scheduling problem is solved for each time-step of the baseline year and is repeated for the remaining years of the planning horizon, while discounting future cash flows to adjust the results for inflation. HOMER Pro is also equipped with the "Multi-Year" module, which allows for modelling the dynamic characteristics of the economic MG planning problem by running a simulation for each ensuing year of the project lifetime. However, since the HOMER Optimizer does not support multi-year planning, first, the optimal capacity configuration was determined using the HOMER Optimizer without considering the dynamic characteristics of the problem, and then the search space was defined based on the preliminary results obtained, with adequate margins, in size steps of 1 unit. Finally, the model was re-run using the HOMER's original grid search algorithm—which simulates all of the feasible equipment capacity combinations with respect to the defined design space. This enabled avoiding sub-optimal solutions, while retaining computational tractability. The conceptualised MG system was simulated under both the cycle-charging and load-following energy dispatch strategies in each simulation case, and the strategy that resulted in a lower levelised cost of energy (LCOE) was selected as the optimum solution.

In addition, HOMER employs specifically-developed algorithms to synthesise one-year of hourly time-series insolation, air temperature, and wind speed data from the corresponding monthly average values to reflect the characteristics of real data (of the same level of granularity) in terms of seasonal, daily, and hourly patterns. Accordingly, hourly-basis, year-round profiles for weather data were derived based on the corresponding monthly average data, shown in Figures 6 and 7.

3.2. Characterisation of Energy Resilience

To quantify the resilience capacity of the MG system to prolonged utility grid outages, the simulation was executed multiple times with varying values of grid reliability. Three built-in parameters in HOMER Pro were used to characterise the grid reliability, namely, (1) mean outage frequency, (2) mean repair time, and (3) repair time variability. Each random outage was injected into the model at a pseudo-random time-step throughout the year-long operational period, while the associated failure duration was determined by independent sampling from a normal distribution defined by mean repair time and repair time variability. Accordingly, the mean outage frequency was varied from 1 to 20 in steps of 1, the mean repair time was varied from 1 h to 168 h (2 weeks) in steps of 4 h, while the repair time variability was fixed at 1 h—which adds a stochastic dimension to the model. That is, 840 ((168 h/4 h) × 20 frequencies) independent outage time-series were produced for the optimisation model. Moreover, to improve the cost-efficiency of the optimal MG capacity configuration solutions, the model was designed to sustain only the critical loads during extended utility grid outages.

In this light, the total load on the MG system was partitioned into critical and non-critical loads, and non-critical loads were turned off when the grid was off. To this end, first, the residential load duration curve was derived by sorting the hourly-basis, one-year residential load curve, as illustrated in Figure 8. Accordingly, the average value of intermediate loads was determined and set as the threshold criterion for residential load partitioning. More specifically, the residential load profile was clipped at this point to form the critical residential load profile. The amounts of residual residential demand that was left from the peak being clipped established the profile of time-stamped non-critical residential load demand, which was only active under the normal, grid-connected operation regime. Additionally, in the grid outage events, the deferrable load was limited to as much as the energy needed to fully charge one EV (Nissan Leaf, battery capacity = 30 kWh [34])—to meet the emergency management needs at the site.

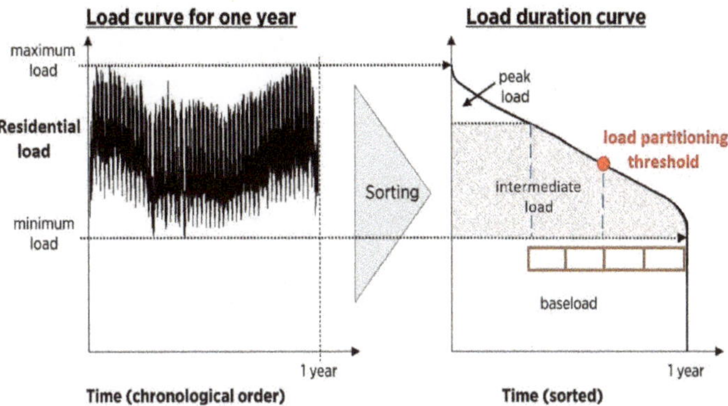

Figure 8. Illustration of the proposed method to determine the load partitioning threshold on a representative load duration curve presented in [62]. (The load curve was taken from the cited reference, and then adapted to describe the proposed load partitioning process.)

3.3. Model Assumptions and Design Standards

Table 5 lists further input data for the model parameters and their respective sources.

Table 5. MG capacity expansion planning model's parameters.

Parameter	Value	Source
Nominal discount rate	4.5%	[63]
Expected inflation rate	1.9%	[64]
Project lifetime	25 years	(this paper)
Minimum autarky ratio [1]	80%	(this paper)
Maximum annual capacity shortage in meeting critical loads	0%	(this paper)
Load growth rate	1.1% per annum	[65]

[1] The time-based autarky ratio indicates the self-sufficiency degree of the MG and can be determined by dividing the yearly number of time-steps where no electricity is purchased from the grid (to meet the full demand) by the total number of time-steps considered for the representative year of the optimisation [66].

Figure 9 shows a flowchart of the research methodology to include energy resilience criteria in the long-term MG optimisation model implemented in HOMER, which is illustrated by means of the considered case study example. As Figure 9 shows, first, the HOMER model is run for each of the above-described 840 outage time-series for the test-case under investigation and the solutions are recorded. Then, the model results of the numerical case example at each simulation run are aggregated to obtain the sensitivity of the total discounted system costs, energy exchange levels with the national grid, and the optimal MG configuration to the variations of the key grid outage parameters.

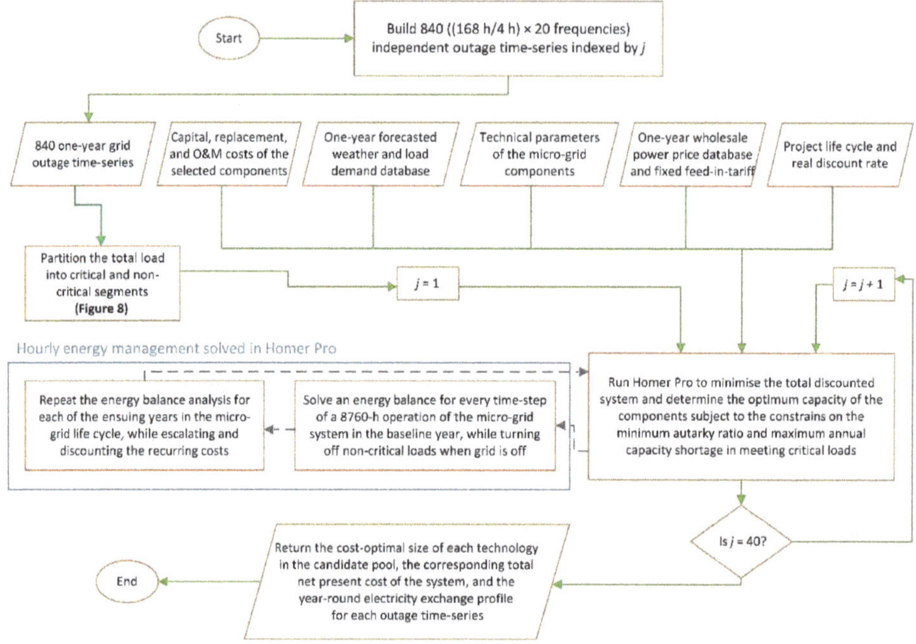

Figure 9. Flowchart of the research methodology.

4. Results and Discussion

This section presents and discusses the numerical simulation results obtained for the test-case MG laid out in Section 2 by applying the method described in Section 3.

4.1. Feasibility and Optimal Capacity Configuration

The optimum equipment capacity configuration was determined first with the grid reliability level set at 100%. Table 6 presents the optimisation results for this baseline scenario. (It should be noted that all the simulation results are based on the manufacturer-provided equipment specifications and technical data, which represent the performance of the equipment under standard test conditions.) The table demonstrates that a WT is a more economically viable power generation technology than solar PV for generation expansion at the considered site, as it could be expected from the historical meteorological data (see Figure 6). However, given the complementary daily and seasonal cycles of wind speed and insolation, the existing installed capacity of solar PV is still expanded in the cost-optimal solution, albeit by a small margin. This not only validates the viability of the existing PV generation system, but also indicates its sizeable impact on the diversification of the feasible future generation mix for the site.

Table 6. Optimal capacity additions and corresponding costs for a 100% reliable grid.

PV [1,2] [kW]	WT [1] [kW]	Battery [1] [kWh]	Inverter [1] [kW]	EVSE [1,3,4] [kW]	TNPC [5] [$]	LCOE [$/kWh]
0.855	4	6.6	3	14.72	35,891	0.094

[1] The optimal equipment capacity was determined as multiples of the selected components' nameplate capacities. [2] The optimal PV capacity represents the newly added capacity of the site's PV plant. [3] The electric vehicle supply equipment (EVSE) did not take part in the optimisation procedure. The optimal size of the EVSE was determined outside the model by finding the total minimum EV charging capacity that meets the peak EV charging demand of the site under cooperative EV charging conditions. The obtained total minimum capacity was then divided by the maximum charging capacity of each charger and rounded up to the nearest integer to obtain the optimal number of chargers. [4] The optimal size of the EVSE is defined based on the maximum charging capacity of each charger. [5] TNPC represents the conceptualised MG's "total net present cost". The reader is referred to [67–69] for a detailed description of the net present value method and levelised cost of energy (LCOE) calculations.

The total net present cost (TNPC) of the system is broken down into the main cost components in Figure 10. Three key observations emerge collectively from Table 6 and Figure 10, which can be generalised to any MG capacity reinforcement problem: (1) the system's low O&M costs helped offset its relatively high capital costs, which were mainly generated by new WT and BESS capacity additions, (2) a comparison of the obtained LCOE with the current average price of domestic electricity in the Wairarapa region, New Zealand, where the site is located ($0.34/kWh [70]) indicates that an optimally planned community-based renewable power generation system even surpasses retail grid parity, and (3) the negative value obtained for the net energy purchased from the utility grid suggests that not only did energy exchange with the utility grid provide power quality benefits and help avoid energy spillage (which was often carried out through a dedicated dump load to balance supply and demand), but it also contributed to the profitability of the project, albeit slightly.

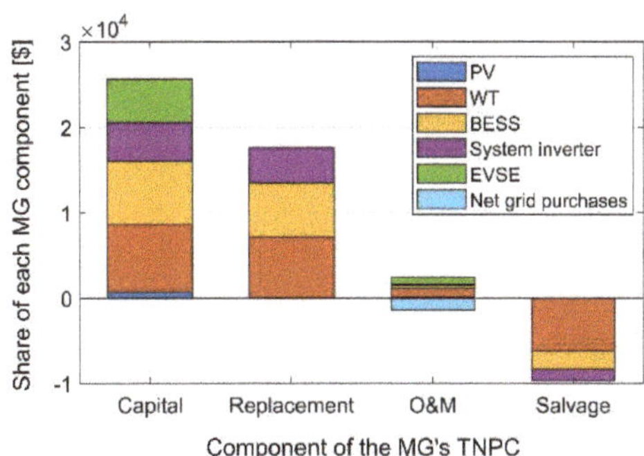

Figure 10. Breakdown of the MG's total net present cost ($).

To understand in more detail the active power flow pattern between the MG and the main grid, the year-round profiles for energy purchased from the grid and energy sold to the grid, obtained for the cost-minimal solution under normal, grid-connected conditions, are plotted in Figures 11 and 12, respectively. The trends visible when comparing the figures are revealing. The battery bank in the optimal equipment combination enabled the MG to engage in energy arbitrage, by charging when utility rates were low and discharging during more remunerative times of day, (The buy-back rate is fixed at NZ$0.08/kWh. However, given the availability of long-term data forecasts, the optimisation protocol incorporates some degree of forward-looking behaviour into the energy exchange decision-making process with respect to the load on the MG in the medium to longer term—identifying the more profitable time-steps for feeding power back into the grid. That is, the profitability is measured independently of the single rate feed-in tariff.) and/or by shifting the excess power from non-dispatchable renewables for sale at more valuable times of day. This, additionally, helped reduce peak demand charges.

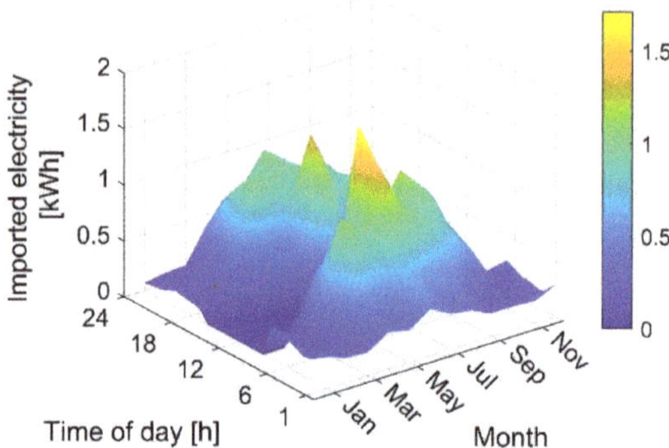

Figure 11. Year-round profile for energy purchased from the grid (kWh). The numerical values belonging to this figure are listed in Table S3 in Supplementary Material.

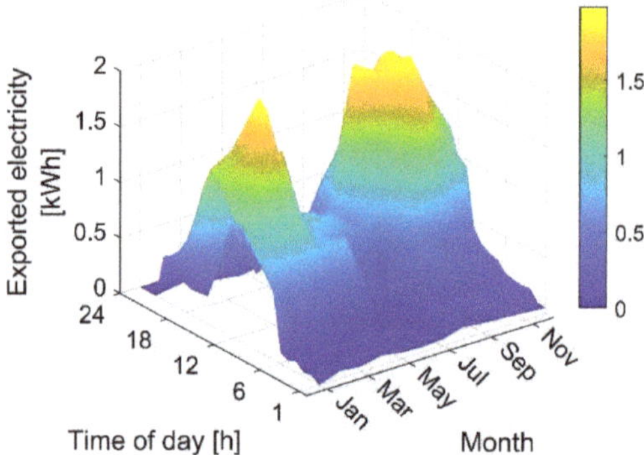

Figure 12. Year-round profile for energy sold to the grid (kWh). The numerical values belonging to this figure are listed in Table S4 in Supplementary Material.

4.2. The Cost of Energy Resilience

Using the methodology described in Section 3 for the planning for energy resilience, the optimal MG capacity expansion problem was solved independently to global optimality for a series of different scenarios defined by mean grid outage frequency and mean grid repair time. As an example, Figure 13 shows the one-year outage database (with hourly intervals) built stochastically for the middle-case scenario, with a grid outage frequency of 10 per year and a mean repair time of 84 h. Accordingly, Figures 14 and 15 respectively depict the resulting TNPC and the corresponding LCOE of the MG expansion under varying degrees of grid reliability characterised by the above two parameters. Moreover, the total annual amount of electricity sold to the grid [kWh] and the total annual amount of electricity purchased from the grid (kWh) at different grid reliability levels are displayed in Figures 16 and 17, respectively. Note that all of the results presented in this sub-section are adjusted for the EVSE costs, which were determined outside the model by matching the charging station's capacity to the peak EV charging demand of the site subject to the cooperative use of the charging infrastructure. Such cooperative behaviour within the community is expected to uphold the installation of only two EV chargers at the site (as indicated in Table 6) to fit for the considered multi-family community's energy demand for mobility.

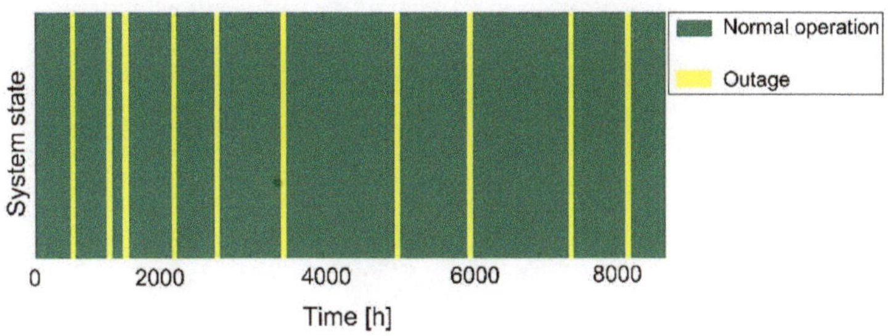

Figure 13. Micro-grid system state (normal/outage) for the middle-case outage scenario.

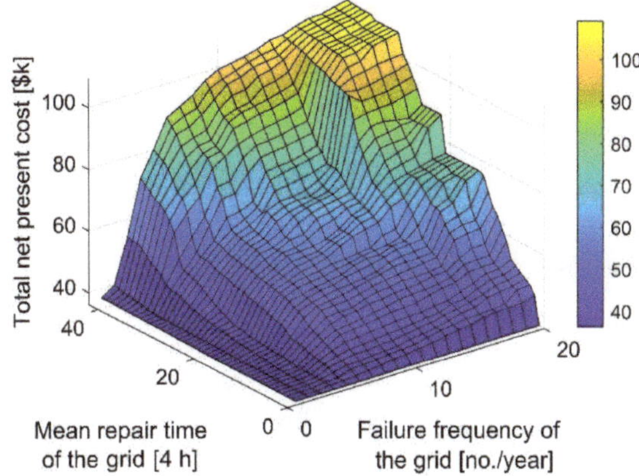

Figure 14. Sensitivity of the MG's TNPC to variations in grid reliability parameters. The numerical values belonging to this figure are listed in Table S5 in Supplementary Material.

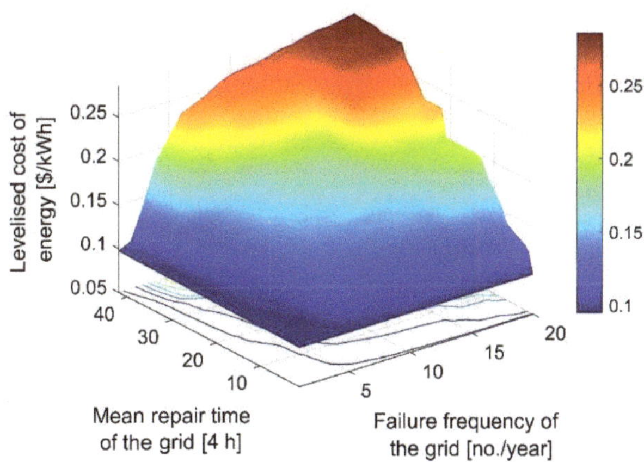

Figure 15. Sensitivity of the MG's LCOE to variations in grid reliability parameters. The numerical values belonging to this figure are listed in Table S6 in Supplementary Material.

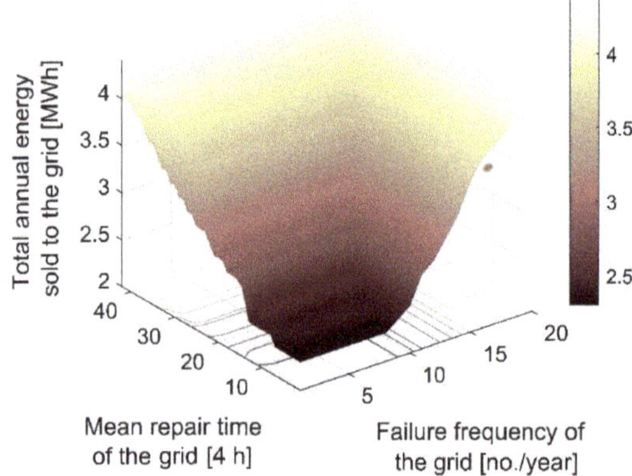

Figure 16. Sensitivity of the total annual energy exports to variations in grid reliability parameters. The numerical values belonging to this figure are listed in Table S7 in Supplementary Material.

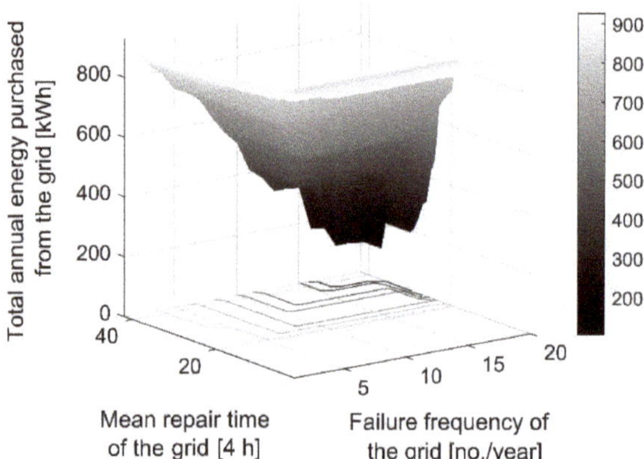

Figure 17. Sensitivity of the total annual energy imports to variations in grid reliability parameters. The numerical values belonging to this figure are listed in Table S8 in Supplementary Material.

The two-way sensitivity analyses, shown in Figures 14–17, are revealing in the following ways:

- The values of total annual energy imports and exports indicate that the MG's net purchased electricity from the grid was approximately a monotonically decreasing function of the grid unreliability. That is, as the failure frequency of the grid and/or its mean repair time increased, the total power sold back to (purchased from) the grid increased (decreases) or remained constant. The underlying reason responsible for this model behaviour is the increase in the excess non-dispatchable power generation capacity of the MG during normal, grid-connected operations, as the grid reliability decreases. However, the increase in revenues generated from trading with the grid, as the MG's resilience to grid outages improved, only partially offset the additional costs incurred. This emphasises the necessity to include a minimum acceptable limit for the target community's energy resilience—to be derived from specifically-developed surveys to estimate the value of lost load—in the resilience-oriented MG capacity expansion planning processes to improve the accuracy of results.

- Non-surprisingly, increasing the unreliability level of the grid through dedicated parameters, increased the TNPC of the MG capacity expansion, which, in turn, increased the LCOE associated with the cost-minimal system.
- The failure frequency of the grid and its mean repair time show almost the same degree of negative effect on the system's life-cycle cost when normalised to the same scale (e.g., in the range of (0 to 1)). Additionally, there seems to exist a front of solutions with respect to the grid reliability parameters, beyond which the whole-life cost of the system grows exponentially. For example, in the middle case scenario, where the grid's mean repair time and failure frequency were respectively considered to be 84 h and 10 per year, the system's TNPC was increased only by about 48%. However, a further 10% increase of either of the above two parameters raised the MG's TNPC by a further 26%. This observation can be rationalised by the change in the MG architecture when the grid unreliability level reaches a critical point, which is discussed in the next sub-section.

4.2.1. Optimal MG System Type

Figure 18 provides the resulting optimal system architecture for different levels of the failure frequency of the grid and the associated mean repair time. As this figure shows, the MG design space is divided approximately equally across two grid-connected system structures, namely, (1) existing PV/added PV/added WT/added BESS, and (2) existing PV/added WT/added BESS. Notably, the figure indicates that it is cost-optimal to add new PV generation capacity for a system resilient to relatively low-frequency, long-duration outages, or high-frequency, short-duration outages, or any combination in between. However, for grid outage scenarios that lie inside the region enclosed by the upper pseudo-triangle shown in Figure 18, the optimally reinforced MG structure excludes additional PV panels. This can be explained by the increase in the probability of occurrence of sustained outages during the night-time hours, where solar panels produce no electricity. Thus, it is more economical to only add WT and BESS capacity, in view of the lower diurnal variations of wind speed and dispatchability of the BESS. It is also worth noting that while the considered bounds for the two-dimensional sensitivity analyses might far exceed the local utility grid's reliability (especially the scenarios that lie in the region enclosed by the upper pseudo-trainable in Figure 18), the results of such sensitivity analyses are important in evaluating the robustness of the baseline system costs and architecture. Accordingly, Figure 18 demonstrates that the obtained grid-connected MG configuration of existing PV/added PV/added WT/added battery in the baseline scenario (100% reliable grid) is highly robust to variations in grid reliability.

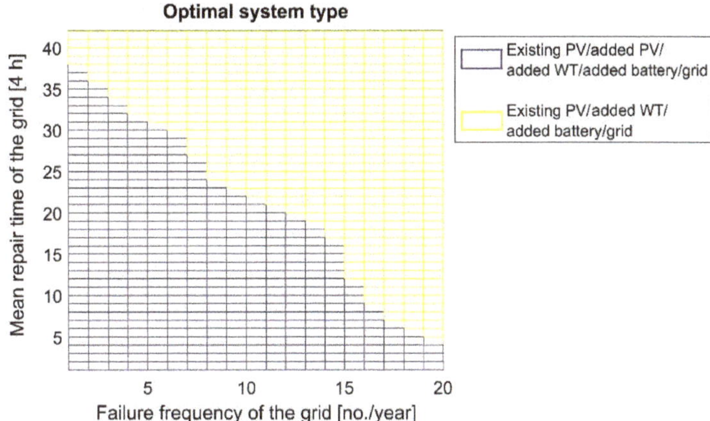

Figure 18. Optimal MG configuration at different grid reliability levels. The numerical values belonging to this figure are listed in Table S9 in Supplementary Material.

4.2.2. Indicative Resilient System Optimisation Analysis

According to the results presented in Figure 14, addressing energy resilience is possible at comparably small total discounted cost increases of about 16% for a sufficiently resilient system (within a New Zealand context), tolerable of two sustained outages per year with each duration of the outage lasting up to four days (or, nearly equally, tolerable of four outages per year each with a maximum outage duration of up to two days), while optimally planning the capacity additions for the site under study. To indicate the effect of planning for energy resilience on the added MG equipment capacity, and in turn, the MG's life-cycle cost, this sub-section details the results obtained for the MG system resilient to two extended outages per year each with a maximum outage duration of up to four days. Moreover, all the analyses presented hereafter are based on this scenario. Table 7 details the capacity expansion planning results obtained for the case considered in this paper under the grid unreliability scenario outlined above. The table demonstrates that the only change in the optimal capacity configuration of the MG with respect to the baseline case (refer to Table 6) is the addition of one more battery module to cater for the prolonged grid outages, which will need to be replaced in year 15 of the project.

Table 7. Optimal capacity additions for the system resilient to 2 extended outages per year each with a maximum duration of up to 4 h.

PV (kW)	WT (kW)	Battery (kWh)	Inverter (kW)	EVSE (kW)	TNPC ($)	LCOE ($/kWh)
0.855	4	9.9	3	14.72	41,783	0.109

4.3. Capital Budgeting Metrics

This section measures the profitability of the investigated MG capacity expansion project based on three key investment appraisal metrics, namely the return on investment (ROI), internal rate of return (IRR), and discounted payback period (DPP). HOMER calculates theses economic sustainability metrics with reference to a base case, which is normally the system with the lowest capital cost that can meet the load demand on the MG. Accordingly, given the presence of the utility grid at the considered site, the base case in this analysis is the existing PV/grid architecture, which incurs no capital costs in view of the presence of a 50 kVA transformer at the site.

4.3.1. Return on Investment

The ROI is defined as the yearly cost savings an investment project generates for the injection of the financial capital, which can be calculated from Equation (7) [71–73]:

$$ROI = \frac{\sum_{i=0}^{PL}\left(CC_{ref}(i) - CC(i)\right)}{PL \times \left(CC - CC_{ref}\right)}, \tag{7}$$

where CC and CC_{ref} represent the capital cost of the suggested system and the reference system, respectively; $CC(i)$ and $CC_{ref}(i)$ denote the nominal cash flow for the proposed system and the reference system in year i; and PL represents the project lifetime.

4.3.2. Internal Rate of Return

Considering the base case defined above, the IRR is defined as the discount rate at which the net present value of all cash flows from the proposed investment equals to that of the base case. Accordingly, the IRR can be calculated as follows [71,72]:

$$IRR = \sum_{i=0}^{PL} \frac{\left(PV(NCI_{ref}(i) - NCI(i)\right)}{(1 + dr)^i} - \left(CC - CC_{ref}\right), \tag{8}$$

where dr is the real discount rate, and the term $(PV(NCI_{ref}(i) - NCI(i)))$ indicates the present value of the difference of the two net cash inflow sequences (for the proposed project and the base case) during the period i, while the net cash inflow represents the system's income from selling electricity to the grid and assets' salvage value minus the components' capital, replacement, and O&M costs, and the costs associated with the purchase of electricity from the grid.

4.3.3. Discounted Payback Period

The DPP refers to the time required to recover the difference in the discounted cash flow difference between the proposed system and the base case system, which can be expressed mathematically as [71,72,74]

$$DPP = \frac{1 - \dfrac{\ln\left(\dfrac{1}{(CC-CC_{ref})\times dr}\right)}{(PV(NCI_{ref}(i)-NCI(i)))}}{\ln(1+dr)},$$
(9)

where ln represents the natural logarithm.

4.3.4. Resulting Cash Flow Metrics

Table 8 lists the obtained cash flow metrics for the proposed MG capacity expansion programme with reference to the case where any shortfalls in meeting the growing residential load power demand and EV charging loads—arising from the shortage in the existing PV generation capacity—is drawn from the utility grid.

Table 8. Calculated capital budgeting metrics for the proposed resilience-oriented, community-scale renewable energy development project.

Metric	Return on Investment (%)	Internal Rate of Return (%)	Discounted Payback Period (Years)
Value	47.63	54.51	4.74

Table 8 demonstrates that the project yields relatively high ROI and IRR values through cost savings realised by building an at least 80% energy self-reliant community energy system, which is able to exploit the differences between the feed-in-tariff rate and wholesale power prices during off-peak periods of energy-use—by making effective use of a battery bank. The table also shows that the conceptualised resilient community MG for critical services would pay for itself in as little as 4.74 years, which provides further support for the financial sustainability of the proposed development plan. Finally, the results of this cost-benefit analysis collectively indicate that not only is the proposed capital project economically feasible without any subsidies delivered as tax incentives (e.g., renewable energy investment tax credits or production tax credits), but it also represents an attractive investment opportunity, which is capable of generating moderate to substantial savings in the community's energy bills in the medium to long run.

5. Conclusions

Radial electricity distribution networks have historically been designed to reduce unplanned, equipment failure-induced outages. However, the combination of multi-day outages as a result of many of the recent extreme weather events, and the improved cost-efficiency of distributed energy resources—including renewable energy generation and storage technologies—have stimulated considerable interest in deploying community-led sustainable energy systems that are capable of meeting critical loads if the upstream network fails. This study developed the first grid-connected community MG equipment capacity expansion planning method that integrates energy resilience constraints without creating redundant capacity. The proposed method yields robust decision-making support by evaluating the trade-off between life-cycle costs and the resilience of energy supply. Specifically, to quantify the resilience of the proposed MG, a time-series of random outages throughout the year were generated for a range of outage frequencies of certain durations and incorporated into the HOMER Pro model.

The numerical simulation results obtained from the application of the model to the case example of Totarabank eco-village, New Zealand, have provided several novel insights:

1. Over the 25-year project life-cycle (planning horizon), the optimally expanded MG system can gain resilience against two outages per year, each up to four days in length, at relatively small discounted cost increases of 16% (equating to NZ$5892). This lends support to the idea that at current costs of renewable energy technologies, it is financially feasible for a community-level site to achieve a sufficient degree of survivability against sustained grid outages.

2. The optimal architecture (component type combination), namely grid-connected existing PV/added PV/added WT/added BESS MG system, determined for the case with a 100% reliable grid, seems to be highly robust against a wide range of grid unreliability values. Much of the reason for this lies in the fact that solar resource tends to complement wind resource at the site. However, the evidence from this study shows that in high wind, low solar regions it is not cost-optimal to add PV capacity when the degree of grid unreliability passes certain limits. This can mainly be attributed to the lower capacity factor of the PV plant.

3. With an IRR of about 55%, the initial investment required for the capacity expansion of the site's energy infrastructure to meet the projected energy consumption growth (driven primarily by the decarbonisation of the transport), can be recouped in less than five years, while additionally ensuring backup power supply to critical loads for two outages per year, each up to four days in length. This, together with the capital affordability of the project, suggests that it can be financed completely by the local community. Moreover, given the demonstrated evidence of the cost-efficiency of such programmes, it is expected to be attractive to many third-party investors.

Although this paper has focused particularly on the capacity expansion planning of grid-tied community MGs, the proposed framework of planning for resilience can be generalised to handle various types of sustainable energy systems of different scales. Moreover, the area of application of the proposed framework is not restricted to the expansion planning problems. It can be easily extended for application to the general renewable energy system capacity planning problem of greenfield sites having access to the national grid.

The study has a number of limitations, namely:

1. All the estimates were based on defining energy resilience in terms of sustained grid outages. That is, the study has not accounted for the impact of extreme weather episodes on the operability of the considered site's renewable energy generation assets.

2. The proposed method does not account for the planned extended power outages related to grid capacity additions or equipment maintenance and repair.

3. The study did not include any inputs from the site on the value the community places on the unserved energy during non-grid-connected operations. It is not implausible that the average annual loads that are deemed critical by end-consumers be different from the adjusted threshold criterion for load partitioning in this study.

4. While the integration of unidirectional EV charging infrastructure is shown to be both technically feasible and economically viable, no attempt is made in this study to investigate the role that bidirectional charging (powered by V2G technology) can play in improving the profitability of the project.

Future Work

To address the limitations outlined above, future work needs to develop a more detailed resilient-oriented MG capacity expansion planning modelling framework that: (1) takes into account the local extreme weather impact on the MG's own onsite renewable energy generation equipment and expands the proposed resilience quantification method to include the hours of the autonomy of the storage bank; (2) integrates scheduled outages into the resilience-oriented MG capacity planning processes by combining them with those of the random unplanned outages to generate an integrated outage time-series to be fed into the optimisation problem; (3) considers the true value of lost load perceived by end-consumers that can be derived from specifically-developed surveys, whilst additionally reflecting the seasonal differences in the elasticity of non-critical loads; and (4) harnesses the potential of EV batteries in facilitating the integration of renewable energy sources, which may provide the opportunity to reduce the total discounted system costs and improve the cost-efficiency of planning for resilience, security, and autonomy of community energy systems.

Supplementary Materials: The following are available online at http://www.mdpi.com/1996-1073/13/15/3970/s1, Table S1: numerical values underlying Figure 4; Table S2: numerical values underlying Figure 5; Table S3: numerical values underlying Figure 11; Table S4: numerical values underlying Figure 12; Table S5: numerical values underlying Figure 14; Table S6: numerical values underlying Figure 15; Table S7: numerical values underlying Figure 16; Table S8: numerical values underlying Figure 17; and Table S9: numerical values underlying Figure 18.

Author Contributions: Conceptualization, S.M.; methodology, S.M.; software, S.M.; validation, S.M., A.C.B., and D.B.; formal analysis, S.M., A.C.B., and D.B.; investigation, S.M.; resources, S.M., A.C.B., and D.B.; data curation, S.M.; writing—original draft preparation, S.M.; writing—review and editing, A.C.B. and D.B.; visualization, S.M.; supervision, A.C.B. and D.B.; and project administration, A.C.B. All authors have read and agreed to the published version of the manuscript.

Funding: This research received no external funding.

Acknowledgments: The authors would like to gratefully acknowledge suggestions and contextual input provided by the Totarabank Subdivision developer Andy Duncan.

Conflicts of Interest: The authors declare no conflict of interest.

Nomenclature

Acronyms

BESS	Battery energy storage system
DPP	Discounted payback period
EV	Electric vehicle
EVSE	Electric vehicle supply equipment
HOMER	Hybrid Optimization of Multiple Energy Resources
IRR	Internal rate of return
LCOE	Levelised cost of energy
MG	Micro-grid
O&M	Operation and maintenance
PV	Photovoltaic
ROI	Return on investment
TNPC	Total net present cost
V2G	Vehicle-to-grid
WT	Wind turbine

Indices

i	Index of year

Scalars

Δt	Length of time-step t in hours
η_{PV}	PV module's efficiency
γ	Wind shear exponent
dr	Real discount rate
DOD	Depth of discharge
$E_{B,min}, E_{B,max}$	Minimum/maximum usable capacity of each battery module

f_{PV}	PV module's derating factor
$G_{T,STC}$	Solar irradiance at standard test conditions
h_{hub}	Hub height of the wind turbine
h_{ref}	Reference height of wind speed records
k_p	PV module's temperature coefficient
$NOCT$	Nominal operating cell temperature
$P_{PV,r}$	PV module's rated power
PL	Project lifetime
T_{STC}	Cell temperature at standard test conditions
Parameters	
P_L^t	Total load power demand on the micro-grid in time t
P_{PV}^t	Power output from the PV system in time t
P_{WT}^t	Power output from the wind turbine system in time t
π_{ex}^t	Wholesale electricity price in time t
FiT	Feed-in-tariff
T_{PV}^t	PV module's temperature in time t
G_T^t	Global horizontal irradiance in time t
T_a^t	Ambient temperature in time t
V_{hub}^t	Hub-height wind speed in time t
Variables	
$cost_g^t$	Cost of power exchange with the main grid
$CC(i)$, $CC_{ref}(i)$	Capital cost of the suggested micro-grid/reference system in year i
E_B^t	Energy content of the battery bank in time t
$NCI(i)$, $NCI_{ref}(i)$	Net cash inflow sequence of the proposed micro-grid/baseline case in year i
P_B^t	Charging/discharging power of the battery bank in time t
P_g^t	Imported/exported power from/to the national grid in time t
Functions	
PV	Present value function

References

1. Warneryd, M.; Håkansson, M.; Karltorp, K. Unpacking the complexity of community microgrids: A review of institutions' roles for development of microgrids. *Renew. Sustain. Energy Rev.* **2020**, *121*, 109690. [CrossRef]
2. Hirsch, A.; Parag, Y.; Guerrero, J. Microgrids: A review of technologies, key drivers, and outstanding issues. *Renew. Sustain. Energy Rev.* **2018**, *90*, 402–411. [CrossRef]
3. Mohseni, S.; Brent, A.C. Smart Grid and Zero-Emissions Energy Systems: The Need for a Multi-Dimensional Investment Planning Perspective. *IEEE Smart Grid eNewsl.* **2018**, *Special Issue on the Paris Climate Agreement and the Role of the Smart Grid.* [CrossRef]
4. Achiluzzi, E.; Kobikrishna, K.; Sivabalan, A.; Sabillon, C.; Venkatesh, B. Optimal Asset Planning for Prosumers Considering Energy Storage and Photovoltaic (PV) Units: A Stochastic Approach. *Energies* **2020**, *13*, 1813. [CrossRef]
5. Chen, T.; Alsafasfeh, Q.; Pourbabak, H.; Su, W. The next-generation US retail electricity market with customers and prosumers—A bibliographical survey. *Energies* **2018**, *11*, 8. [CrossRef]
6. Mohseni, S.; Brent, A.C.; Burmester, D. The Role of Artificial Intelligence in the Transition from Conventional Power Systems to Modernized Smart Grids. *IEEE Smart Grid eNewsl.* **2019**, *Special Issue on Grid Modernization.* [CrossRef]
7. Anderson, K.H.; DiOrio, N.A.; Cutler, D.S.; Butt, R.S. Increasing resiliency through renewable energy microgrids. *J. Energy Manag.* **2017**, *2*, 2.
8. Eskandarpour, R.; Lotfi, H.; Khodaei, A. Optimal microgrid placement for enhancing power system resilience in response to weather events. In Proceedings of the 2016 North American Power Symposium (NAPS), Denver, CO, USA, 18–20 September 2016; pp. 1–6.
9. Barnes, A.; Nagarajan, H.; Yamangil, E.; Bent, R.; Backhaus, S. Tools for improving resilience of electric distribution systems with networked microgrids. *arXiv* **2017**, arXiv:1705.08229.

10. Hussain, A.; Bui, V.-H.; Kim, H.-M. Microgrids as a resilience resource and strategies used by microgrids for enhancing resilience. *Appl. Energy* **2019**, *240*, 56–72. [CrossRef]

11. Boloukat, M.H.S.; Foroud, A.A. Stochastic-based resource expansion planning for a grid-connected microgrid using interval linear programming. *Energy* **2016**, *113*, 776–787. [CrossRef]

12. Waqar, A.; Wang, S.; Dawoud, S.M.; Tao, T.; Wang, Y. Optimal capacity expansion-planning of distributed generation in microgrids considering uncertainties. In Proceedings of the 2015 5th International Conference on Electric Utility Deregulation and Restructuring and Power Technologies (DRPT), Changsha, China, 26–29 November 2015; pp. 437–442.

13. Motie, S.; Keynia, F.; Ranjbar, M.R.; Maleki, A. Generation expansion planning by considering energy-efficiency programs in a competitive environment. *Int. J. Electr. Power Energy Syst.* **2016**, *80*, 109–118. [CrossRef]

14. Hemmati, R.; Saboori, H.; Siano, P. Coordinated short-term scheduling and long-term expansion planning in microgrids incorporating renewable energy resources and energy storage systems. *Energy* **2017**, *134*, 699–708. [CrossRef]

15. He, Y.; Sharma, R. Microgrid generation expansion planning using agent-based simulation. In Proceedings of the 2013 IEEE PES Innovative Smart Grid Technologies Conference (ISGT), Washington, DC, USA, 24–27 February 2013; pp. 1–6.

16. Salimi, M.; Nasr, M.-A.; Hosseinian, S.H.; Gharehpetian, G.B.; Shahidehpour, M. Information Gap Decision Theory-Based Active Distribution System Planning for Resilience Enhancement. *IEEE Trans. Smart Grid* **2020**, (accepted; in press). [CrossRef]

17. Ibanez, E.; Lavrenz, S.; Gkritza, K.; Mejía, D.; Krishnan, V.; McCalley, J.; Somani, A.K. Resilience and robustness in long-term planning of the national energy and transportation system. *Int. J. Crit. Infrastruct.* **2016**, *12*, 82. [CrossRef]

18. Shao, C.; Shahidehpour, M.; Wang, X.; Wang, X.; Wang, B. Integrated planning of electricity and natural gas transportation systems for enhancing the power grid resilience. *IEEE Trans. Power Syst.* **2017**, *32*, 4418–4429. [CrossRef]

19. Celtic Energy. *Resilient Microgrids for Rhode Island Critical Services*; Celtic Energy: Caerphilly, UK, 2017.

20. Yuan, W.; Wang, J.; Qiu, F.; Chen, C.; Kang, C.; Zeng, B. Robust optimization-based resilient distribution network planning against natural disasters. *IEEE Trans. Smart Grid* **2016**, *7*, 2817–2826. [CrossRef]

21. Wu, X.; Wang, Z.; Ding, T.; Wang, X.; Li, Z.; Li, F. Microgrid planning considering the resilience against contingencies. *IET Gener. Transm. Distrib.* **2019**, *13*, 3534–3548. [CrossRef]

22. Dong, J.; Zhu, L.; Su, Y.; Ma, Y.; Liu, Y.; Wang, F.; Tolbert, L.M.; Glass, J.; Bruce, L. Battery and backup generator sizing for a resilient microgrid under stochastic extreme events. *IET Gener. Transm. Distrib.* **2018**, *12*, 4443–4450. [CrossRef]

23. Totarabank–Sustainable Rural Living. Available online: https://totarabank.weebly.com/ (accessed on 25 May 2020).

24. Battery Systems. Available online: https://www.nexeon.co.uk/technology/ (accessed on 25 May 2020).

25. Prasad, D.; Snow, M. *Designing with Solar Power: A Source Book for Building Integrated Photovoltaics (BIPV)*; Routledge: Abingdon-on-Thames, UK, 2014; ISBN 1134032501.

26. The X2000L Wind Turbine System. Available online: http://www.solazone.com.au/1kw-2kw-wind-turbines/ (accessed on 25 May 2020).

27. Gipe, P. Wind power. *Wind Eng.* **2004**, *28*, 629–631. [CrossRef]

28. Mohseni, S.; Brent, A.; Burmester, D.; Chatterjee, A. Optimal Sizing of an Islanded Micro-Grid Using Meta-Heuristic Optimization Algorithms Considering Demand-Side Management. In Proceedings of the 2018 Australasian Universities Power Engineering Conference (AUPEC), 2018, Auckland, New Zealand, 27–30 November 2018; pp. 1–6. [CrossRef]

29. Moghaddas-Tafreshi, S.M.; Jafari, M.; Mohseni, S.; Kelly, S. Optimal operation of an energy hub considering the uncertainty associated with the power consumption of plug-in hybrid electric vehicles using information gap decision theory. *Int. J. Electr. Power Energy Syst.* **2019**, *112*, 92–108. [CrossRef]

30. Moghaddas-Tafreshi, S.M.; Mohseni, S.; Karami, M.E.; Kelly, S. Optimal energy management of a grid-connected multiple energy carrier micro-grid. *Appl. Therm. Eng.* **2019**, *152*, 796–806. [CrossRef]

31. Yilmaz, M.; Krein, P.T. Review of battery charger topologies, charging power levels, and infrastructure for plug-in electric and hybrid vehicles. *IEEE Trans. Power Electron.* **2012**, *28*, 2151–2169. [CrossRef]

32. 2014 Nissan Leaf. Available online: https://www.thecarconnection.com/overview/nissan_leaf_2014/ (accessed on 25 May 2020).

33. Ode to the Nissan Leaf. Available online: http://evtalk.co.nz/ode-to-the-nissan-leaf/ (accessed on 25 May 2020).

34. 2014 Nissan Leaf review: Electric car landmark. Available online: https://uk.motor1.com/reviews/139275/2014-nissan-leaf/ (accessed on 25 May 2020).

35. Mohseni, S.; Moghaddas-Tafreshi, S.M. A multi-agent system for optimal sizing of a cooperative self-sustainable multi-carrier microgrid. *Sustain. Cities Soc.* **2018**, *38*, 452–465. [CrossRef]

36. Kempton, W.; Tomić, J. Vehicle-to-grid power fundamentals: Calculating capacity and net revenue. *J. Power Sources* **2005**, *144*, 268–279. [CrossRef]

37. Trina Datasheet. Available online: https://static.trinasolar.com/sites/default/files/EN_TSM_PD05_datasheet_B_2017_web.pdf (accessed on 25 May 2020).

38. Renew. Solar Panels. Available online: https://renew.org.au/wp-content/uploads/2019/12/150_solar_systems_guide_table.pdf (accessed on 25 May 2020).

39. RESU: Residential Energy Storage Unit for Photovoltaic Systems. Available online: https://solarjuice.com.au/wp-content/uploads/2016/11/RESU-48V_installation_manual_R2_eng_2016-10-21-1.pdf (accessed on 25 May 2020).

40. Change Your Energy Charge Your Life. Available online: https://solarjuice.com.au/wp-content/uploads/2016/10/160712_New-RESU-leaflet_Global-small.pdf (accessed on 25 May 2020).

41. Battery Energy Storage Systems. Prices for the LG Chem RESU 3.3 kWh. Available online: http://www.batteryenergystoragesystems.com.au/ (accessed on 25 May 2020).

42. SP PRO AU Series –Selectronic. Available online: http://www.selectronic.com.au/brochure/BR0007_09%20SP%20PRO%20Series%20II%20Data%20Sheet%20Web.pdf (accessed on 25 May 2020).

43. Marshall Solar & Energy. Available online: https://marshallsolarandenergy.com.au/product/selectronic-sp-pro-inverter-charger-3-0-kw-24v-spmc240-au/ (accessed on 25 May 2020).

44. Mohseni, S.; Brent, A.C.; Burmester, D. A demand response-centred approach to the long-term equipment capacity planning of grid-independent micro-grids optimized by the moth-flame optimization algorithm. *Energy Convers. Manag.* **2019**, *200*, 112105. [CrossRef]

45. Mohseni, S.; Brent, A.C.; Burmester, D. A comparison of metaheuristics for the optimal capacity planning of an isolated, battery-less, hydrogen-based micro-grid. *Appl. Energy* **2020**, *259*, 114224. [CrossRef]

46. Mohseni, S.; Brent, A.C. Economic viability assessment of sustainable hydrogen production, storage, and utilisation technologies integrated into on-and off-grid micro-grids: A performance comparison of different meta-heuristics. *Int. J. Hydrogen Energy* **2019**, (accepted; in press). [CrossRef]

47. Nikolaus, R.; Paul, T.J. Feasibility Study of Small and Micro Wind Turbines for Residential Use in New Zealand: An Analysis of Technical Implementation, Spatial Planning Processes and of Economic Viability of Small and Micro Scale Wind Energy Generation Systems for Residential Use in New Zealand. Available online: https://researcharchive.lincoln.ac.nz/handle/10182/4141 (accessed on 25 May 2020).

48. Mohseni, S.; Brent, A.C.; Burmester, D. A Sustainable Energy Investment Planning Model Based on the Micro-Grid Concept Using Recent Metaheuristic Optimization Algorithms. In Proceedings of the 2019 IEEE Congress on Evolutionary Computation (CEC), Wellington, New Zealand, 10–13 June 2019; pp. 219–226. [CrossRef]

49. Chargers Direct. Delta Mini AC Electric Vehicle Charge Point-Networked 3G. Available online: https://www.chargersdirect.com.au/product/delta-ac-mini-plus/ (accessed on 25 May 2020).

50. Delta AC Mini Plus EV Charger. Available online: https://yhipower.co.nz/catalog/ac-mini-plus-104892.htmx (accessed on 25 May 2020).

51. Elkadeem, M.R.; Wang, S.; Azmy, A.M.; Atiya, E.G.; Ullah, Z.; Sharshir, S.W. A systematic decision-making approach for planning and assessment of hybrid renewable energy-based microgrid with techno-economic optimization: A case study on an urban community in Egypt. *Sustain. Cities Soc.* **2020**, *54*, 102013. [CrossRef]

52. NASA Surface Meteorology and Solar Energy database. Available online: https://eosweb.larc.nasa.gov/ (accessed on 25 May 2020).

53. Solar Power Buy-Back Rates. Available online: https://www.mysolarquotes.co.nz/about-solar-power/residential/solar-power-buy-back-rates-nz/ (accessed on 25 May 2020).

54. Bücher, K. Site dependence of the energy collection of PV modules. *Sol. Energy Mater. Sol. Cells* **1997**, *47*, 85–94. [CrossRef]

55. Mohseni, S.; Brent, A.C.; Burmester, D.; Chatterjee, A. Stochastic Optimal Sizing of Micro-Grids Using the Moth-Flame Optimization Algorithm. In Proceedings of the 2019 IEEE Power & Energy Society General Meeting (PESGM), Atlanta, GA, USA, 4–8 August 2019; pp. 1–5. [CrossRef]

56. Wind Shear. Available online: https://www.engineeringtoolbox.com/wind-shear-d_1215.html/ (accessed on 25 May 2020).

57. The Electricity Market Information: The New Zealand Electricity Authority's Wholesale Database. Available online: https://www.emi.ea.govt.nz/Wholesale/Reports/ (accessed on 25 May 2020).

58. Anderson, B.; Eyers, D.; Ford, R.; Ocampo, D.G.; Peniamina, R.; Stephenson, J.; Suomalainen, K.; Wilcocks, L.; Jack, M. *New Zealand GREEN Grid Household Electricity Demand Study 2014–2018*; UK Data Service: Colchester, UK, 2018. [CrossRef]

59. Anderson, B. *NZ GREEN Grid Household Electricity Demand Data: EECA Data Analysis (Part. C) Upscaling Advice Report v1.0*; University of Otago: Dunedin, New Zealand, 2019.

60. Twidell, J.; Weir, T. *Renewable Energy Resources*; Routledge: Abingdon-on-Thames, UK, 2015; ISBN 1317660374.

61. *HOMER Pro*; HOMER Energy LLC: Boulder, CO, USA, 2020.

62. Ueckerdt, F.; Kempener, R. *From Baseload to Peak: Renewables Provide a Reliable Solution*. International Renewable Energy Agency (IRENA): Abu Dhabi, UAE, 2015.

63. Discount Rates and CPI Assumptions for Accounting Valuation Purposes. Available online: https://treasury.govt.nz/information-and-services/state-sector-leadership/guidance/financial-reporting-policies-and-guidance/discount-rates/discount-rates-and-cpi-assumptions-accounting-valuation-purposes/ (accessed on 25 May 2020).

64. New Zealand Inflation Rate. Available online: https://tradingeconomics.com/new-zealand/inflation-cpi/ (accessed on 25 May 2020).

65. New Zealand's Energy Outlook: Electricity Insight. Available online: https://www.mbie.govt.nz/building-and-energy/energy-and-natural-resources/energy-statistics-and-modelling/energy-modelling/new-zealands-energy-outlook/new-zealands-energy-outlook-electricity-insight/ (accessed on 25 May 2020).

66. Wanitschke, A.; Pieniak, N.; Schaller, F. Economic and environmental cost of self-sufficiency-analysis of an urban micro grid. *Energy Procedia* **2017**, *135*, 445–451. [CrossRef]

67. Mohseni, S.; Brent, A.C.; Burmester, D. A Reliability-Oriented Cost Optimisation Method for Capacity Planning of a Multi-Carrier Micro-Grid: A Case Study of Stewart Island, New Zealand. *arXiv* **2019**, arXiv:1906.09544.

68. Mohseni, S.; Moghaddas-Tafreshi, S.M. Development of a multi-agent system for optimal sizing of a commercial complex microgrid. *arXiv* **2018**, arXiv:1811.12553.

69. Mohseni, S.; Moghaddas-Tafreshi, S.M. A multi-agent approach to optimal sizing of a combined heating and power microgrid. *arXiv* **2018**, arXiv:1812.11076.

70. Electricity Cost and Price Monitoring. Available online: https://www.mbie.govt.nz/building-and-energy/energy-and-natural-resources/energy-statistics-and-modelling/energy-statistics/energy-prices/electricity-cost-and-price-monitoring/ (accessed on 25 May 2020).

71. Seitz, N.; Ellison, M. *Capital Budgeting and Long-Term Financing Decisions*; Harcourt Brace College Publishers: New York, NY, USA, 1995; ISBN 0030754771.

72. Compare Economics—HOMER Energy. Available online: https://www.homerenergy.com/products/pro/docs/latest/compare_economics.html/ (accessed on 25 May 2020).

73. Return on Investment—HOMER Energy. Available online: https://www.homerenergy.com/products/pro/docs/latest/return_on_investment.html/ (accessed on 25 May 2020).

74. Fisher, I. *The theory of Interest, as Determined by Impatience to Spend Income and Opportunity to Invest It*; Macmillan: New York, NY, USA, 1930.

Article

Study on the Suitability of Passive Energy in Public Institutions in China

Shui Yu *, He Liu, Lu Bai and Fuhong Han

School of Municipal and Environmental Engineering, Shenyang Jianzhu University, Shenyang 110168, China;
liuhe@stu.sjzu.edu.cn (H.L.); JJL@stu.sjzu.edu.cn (L.B.); Hanfuhong1995@stu.sjzu.edu.cn (F.H.)
* Correspondence: hj_yushui@sjzu.edu.cn; Tel.: +86-188-0248-3528

Received: 23 May 2019; Accepted: 19 June 2019; Published: 25 June 2019

Abstract: To solve the problem of the low utilization ratio of clean renewable energy in public institutions, the basic information of energy utilization in public institutions was investigated. The suitability of passive energy use in public institutions was studied. According to the basic information and evaluation index of passive energy utilization in public institutions, the suitability of different types of passive energy (solar and geothermal energy) was studied by combining the resource conditions in different climate zones and the characteristics of energy utilization in typical public institutions, and the suitability distribution map was formed. In terms of research methods, the CRITIC (Criteria Importance Though Intercrieria Correlation) method based on the characteristics of objective data, and the natural breakpoint method based on the structure of objective data, were selected. Based on the climatic zoning of the buildings, this study conducted a suitability zoning. Each climatic region of buildings was divided into three sub-regions, which were the passive energy suitability regions of public institutions in the climatic region of the buildings. Finally, all of the regions with the same suitability were partitioned in order to obtain the final results. The distribution map of suitability for the different types of public institution buildings in the different regions is creatively established, which provides the basis for the selection of passive energy technology application schemes for public institution buildings in different regions, and provides macro guidance for energy planners and scheme designers.

Keywords: public institution; passive energy; CRITIC method

---

1. Background

After more than ten years of development, with the gradual attention of the state to the energy conservation work of public institutions, public institutions in various places have carried out many explorations in their own energy management and have achieved certain results [1–7] The relevant systems research and construction have been carried out throughout the country [8]. Some provinces and cities have introduced local approaches and supporting policies. In Beijing, solar hot water systems are mandatory in new homes. In Fujian, measures have been taken to strengthen the promotion, application, and management of renewable energy in civil buildings. Guangdong has incorporated green building management requirements into legislation. Some provinces and cities have carried out energy consumption data statistics, energy audits, or publicity for some public institutions. At the same time, however, there are also some problems that restrict the continuous promotion of the establishment of energy-saving institutions, and have not yet played a role in social demonstration.

In 2017, China accounted for 23.2% of the global energy consumption and 33.6% of the global energy consumption growth. China has led the world in energy growth for 17 consecutive years. In 2017, China's natural gas consumption increased by 15%, accounting for 32.6% of the net increase in global natural gas consumption. Coal accounted for 60.4% of China's energy mix in 2017, down from the 73.6% of a decade earlier and the 62.0% of 2016 [9]. The five-year plan is an important part of China's national economic plan, as well as its long-term plan. It mainly anticipates major national construction projects, the distribution of productive forces, and a significant proportion of the national economy, and sets goals and directions for the long-term development of the national economy. The outline of the 12th Five-Year Plan for the National Economic and Social Development of the People's Republic of China (2011–2015) is referred to as the 12th Five-Year Plan. The energy consumption structure of public institutions in the 12th Five-Year Plan is as follows: electricity accounts for 45.37%, raw coal 30.86%, and other 23.77%. Compared with 2010, the proportion of electricity increased by 11.07 percentage points, while that of raw coal decreased by 17.16 percentage points. Compared with 2012, the per capita energy consumption of the national public institutions decreased by 15.21% in 2017, and the energy consumption per unit of building area decreased by 12.39%. The average energy consumption decreased by 12.85% in 2016, and the energy consumption of construction area decreased by 10.17%. Compared with 2008, the energy consumption per unit building area of public institutions decreased by 22.9% in 2017, and the per capita energy consumption decreased by 29.4%.

However, as of the 12th Five-Year Plan, China's total building energy consumption was 857 million tons of standard coal, accounting for 20% of the country's total energy consumption. The total energy consumption of public institutions is 183 million tons of standard coal, accounting for 21.35% of the total energy consumption of buildings in China, and 4.26% of the total energy consumption of the whole society. The per capita comprehensive energy consumption is 370.73 kg of standard coal per person, and the energy consumption per unit building area is 20.55 kg of standard coal per square meter. The energy consumption per unit area of urban residential buildings is 12.90 kg of standard coal per square meter, and that of the rural residential buildings is 7.82 kg of standard coal per square meter. The energy consumption per unit floor area of public institutions is 1.60 times that of urban residential buildings, and 2.63 times that of rural residential buildings [10].

Donglin Zhang, of Chongqing University, points out that the government's focus on energy efficiency in office buildings is to reduce the use of electricity and increase the use of renewable energy [11]. The energy intensity of office buildings of state organs in China is high, which is an abnormal situation. Meanwhile, with the development of residential buildings to high-rise and super-high-rise buildings in recent years, the energy consumption of power facilities has increased significantly, while the utilization rate of solar energy has decreased significantly, leading to the increase of energy intensity of such buildings [12]. Moreover, the utilization rate of clean renewable energy in universities and hospitals is low [13].

To sum up, public institutions are not only an important subject of energy consumption, but also an important field of energy conservation. This study explores and studies the current status, existing problems, and causes of passive energy applications, and promotes the use of passive energy sources in public institutions in China. In view of the current low utilization rate of clean renewable energy in public institutions, the basic information and research about the energy utilization of public institutions was carried out in order to study the suitability of passive energy utilization in public institutions, and to creatively establish the buildings suitable for different regions and different types of public institutions. The suitability map provides the basis for the selection of passive energy technology application schemes for public buildings in situations in different regions, providing macro guidance for energy planning and solution designers.

2. Content

Passive energy: For buildings, renewable energy that can be used in large quantities is a passive energy source. Renewable energy is inexhaustible, but not all renewable energy can be used for buildings. So, renewable energy includes passive energy, and solar energy, geothermal energy, and wind energy are all passive energy sources. The passive energy sources in the study are solar energy and geothermal energy. Active energy: For buildings, non-renewable energy, secondary energy, and renewable energy that cannot be used in large quantities are active energy sources, such as natural gas (non-renewable energy) and electric energy (secondary energy).

China is located in the eastern part of Eurasia in the Northern Hemisphere, mainly in the temperate zone and subtropical zone, with relatively abundant solar energy resources. According to the long-term observation data accumulated by more than 700 meteorological stations nationwide, the annual total solar radiation in China is roughly between 3.35×103 MJ/m^2 and 8.40×103 MJ/m^2, with an average value of about 5.86×103 MJ/m^2. China is rich in geothermal resources [14]. More than 2700 geothermal outcrops have been discovered. Most provinces (districts) in China have different geothermal outcrops, and there are many hotspots in Yunnan—345 places in Tibet, 342 places in Tibet, 320 places in Hebei, 295 places in Sichuan, and 229 in Guangdong, and so on. Most of the geothermal resources in China are in the low temperatures and have hot water, and there are only 600 hot spots above 80 °C. From the perspective of geothermal distribution in China, there is a trend of increasing geothermal quantity and increasing water temperature from the central to the eastern continental margin and the southwest. According to the survey and evaluation results of the China Geological Survey of the Ministry of Land and Resources, the amount of shallow geothermal energy resources in the cities above the prefecture level is equivalent to 95×108 t of standard coal, and the annual recoverable resources are equivalent to 7×108 t of standard coal. China has many types of geothermal resources, a wide distribution, large reserves, and great potential for development and utilization.

In view of the lack of basic data for passive energy applications in public institutions and the underutilization of passive energy sources, the scope and objectives of the research are first determined. The utilization of solar energy in public institutions mainly includes photothermal conversion, photovoltaic power generation, and solar lighting, and the utilization of geothermal energy in public institutions mainly includes underground water source heat pumps and soil source heat pumps. The main types of public institutions are government agencies (mainly office buildings) and educational buildings (for teaching in universities). In the three aspects of building and health-related buildings (mainly in outpatient and inpatient buildings), passive energy mainly considers solar energy and shallow geothermal energy. The utilization of solar energy is photothermal conversion, and the utilization of geothermal energy is shallow geothermal energy.

Public institutions were referred to as "government agencies" in early comprehensive policy documents. Among the ten key energy-saving projects in the "11th Five-Year Plan", the "Government Institutions Energy Conservation Project" defines government agencies as government agencies, institutions, and social organizations at all levels, including the military, armed police, education, public services, and so on. Government agencies include party organs, people's organs, administrative organs, Chinese People's Political Consultative Conference organs, judicial organs, procuratorial organs, and other public institutions, such as those directly under the state organs, and receiving all or part of financial funds for education, science and technology, culture, health, sports, and other related public welfare industries. Business unit organizations include social groups and related organizations such as workers, youth, and women who use financial funds in whole or in part. In the Regulations on Energy Conservation of Public Institutions promulgated in October 2008, public institutions are defined as state organs, institutions, and organizations that use financial funds in total or in part [15].

2.1. Government Buildings

Government buildings are public buildings constructed by central or local governments at all levels through financial allocation or administrative financing, and they are exclusively used by government departments at all levels. Government office buildings are mainly used for public management affairs and public services. The social functions they carry are necessary for social and economic development. The distribution of building energy consumption aims to meet the reasonable-use functions of buildings [11]. (1) Functional features of government office buildings: (a) Office space, including the leadership office and the general staff office. (b) Public service rooms, including conference rooms, reception rooms, printing rooms, archives, information centers, mailrooms, guard rooms, toilets, and so on. (c) Equipment room, including refrigeration machine room, variable pressure pump room, elevator machine room, power distribution room, and so on. (d) Auxiliary rooms, including canteens, fire rooms, garages, and so on. (2) Government building features: Government buildings are mostly multi-story buildings with sloping roofs or high-rise buildings with flat roofs. Limited by decoration and energy-saving standards, few government office buildings are equipped with glass curtain walls. The exterior is solemn and there is no balcony [16].

The factors affecting the energy consumption of government office buildings are as follows: (1) Construction scale. Buildings larger than 20,000 m^2 are classified as large government buildings, otherwise, they are classified as non-large government buildings. According to the survey and analysis, in non-large government office buildings, the energy consumption generally decreases with the increase in the building scale, but it shows a smooth trend in a certain interval. In general, the energy consumption per unit area of large government office buildings is much lower than that of non-large government office buildings, as shown in Figures 1 and 2.

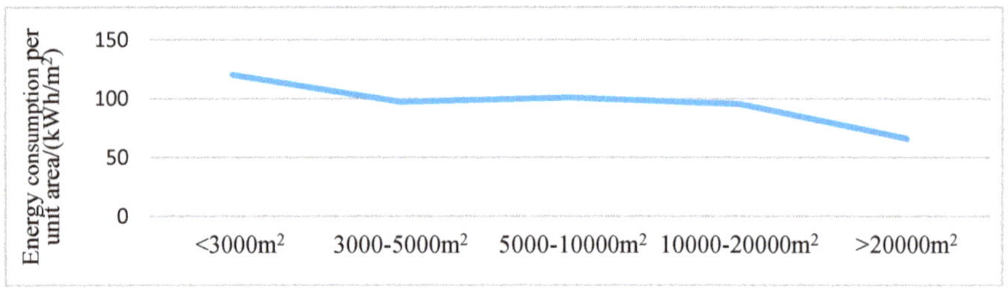

Figure 1. Relationship between the scale and energy consumption of non-large government office buildings.

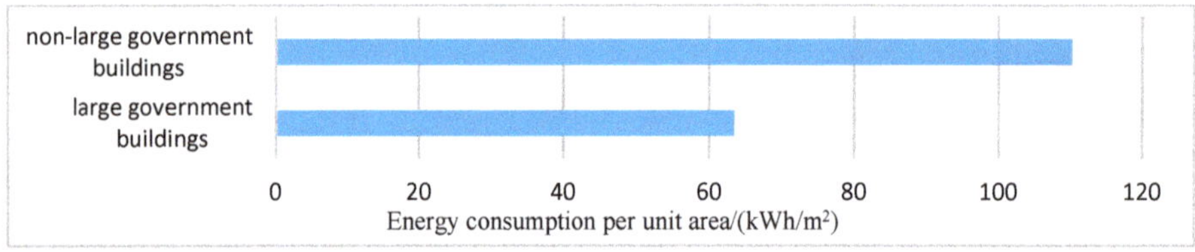

Figure 2. Energy consumption map of different building areas of government office buildings.

(2) Air conditioning form. The survey found that the energy consumption of government office buildings is easily affected by personnel behavior and load rate [17]. Therefore, although the central air conditioning system is superior to the non-central air conditioner in the performance of a single unit, the application process is not for a central air conditioning system to save energy and the central air conditioning system [11], as shown in Figure 3.

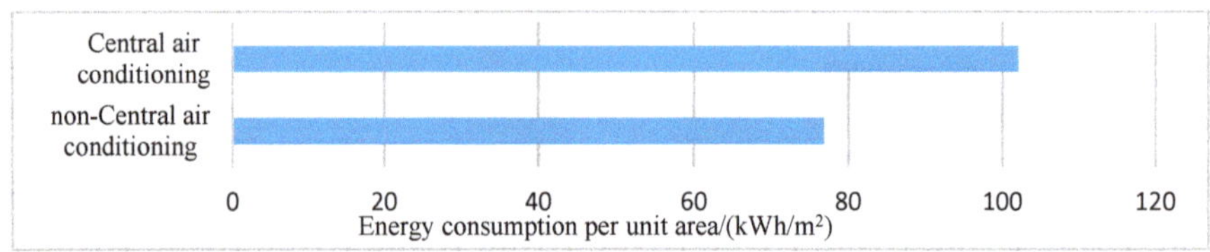

Figure 3. Energy consumption per unit area of different air conditioning forms in government office buildings.

2.2. University Building

A university building is not a building, but a building group that contains many types of buildings. The functions of various buildings are not identical. The main types of college buildings are as follows: (a) Laboratory building. Most of the equipment used in the experimental building is mostly used during the day, and its energy consumption is closely related to various professions. Therefore, the energy consumption of the experimental building is significantly related to the type of colleges and universities. (b) Administrative building. The administrative building is mainly an office space. The density of the energy consumption is related to the working hours. The energy consumption is related to its scale. The site has high requirements for environmental comfort. The air conditioning energy consumption is large in summer, and the energy consumption of indoor office equipment is relatively fixed. The frequency of use of power equipment, such as elevators, is high during commuting hours. (c) Teaching building. The teaching building is a place for students to study and attend classes. As a result of the concentration of personnel, the main energy-consuming equipment in winter is certain lighting fixtures and multimedia. In summer, the air-conditioning energy consumption is too large, and the utilization rate of night classrooms affects the total energy consumption of such buildings.

From what has been discussed above, the sum of the energy consumption of different functional buildings in colleges and universities constitutes the total energy consumption of campus buildings. The energy consumption of public buildings in colleges and universities is related to many factors, such as the climate zone in which universities are located, as well as the types of colleges and universities. However, each university building group includes seven categories, namely: laboratory buildings, administrative buildings, teaching buildings, gymnasiums, libraries, dormitories, and restaurants. Common public buildings (because dormitories and canteens are gradually contracted by enterprises, so they are not in the scope of public institutions): the energy consumption of these buildings accounts for a large proportion of the total energy consumption of the campus [18], and the research is mainly based on teaching buildings.

2.3. Hospital Building

The outpatient department is the first place to contact the patient. The patient and family members must come to the hospital first to register in the clinic, and then will distribute to the various parts of the hospital. Therefore, the outpatient building is the place with the highest population density. Hospital composition: According to the functional department, it can be divided into various outpatient departments, public departments, and medical technology departments. The outpatient department includes the departments of internal medicine, surgery, pediatrics, office, treatment room, and so on. The public department includes the registration office, the toll collection office, the checkout office, and the medicine taking office, and the medical technology department includes the laboratory and the X-ray inspection room. Characteristics: (1) The flow density is large, and the stagnation time is extremely long. (2) The service target is a special group, so the air environment quality in the ward is very high. (3) The air conditioning system is a seasonal air conditioning system. (4) There is a certain pressure difference between the inside of the clinic and the corridor, so as to maintain the positive pressure in the room [19]. (5) Seasonal characteristics: The survey found that the seasonal characteristics of hospital energy consumption are significant. The hospital's electricity consumption is different in different months. Electricity peaks occur in July–September, while other seasons use less electricity, as shown in Figure 4. (6) Regional characteristics: Because the number of hospitals and climate characteristics are different in different regions, the energy consumption varies greatly, as shown in Figure 5 [20].

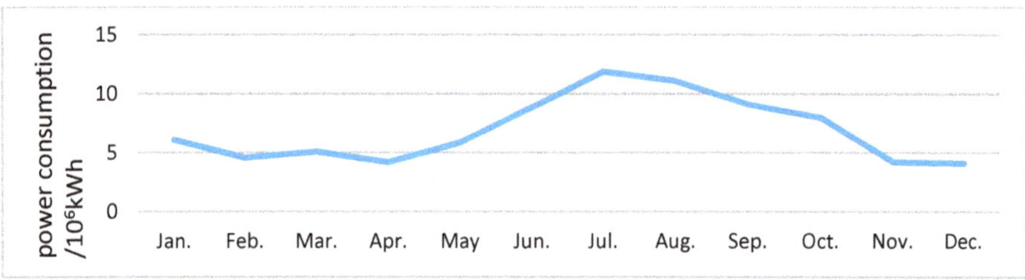

Figure 4. Monthly power consumption-change chart of a large hospital.

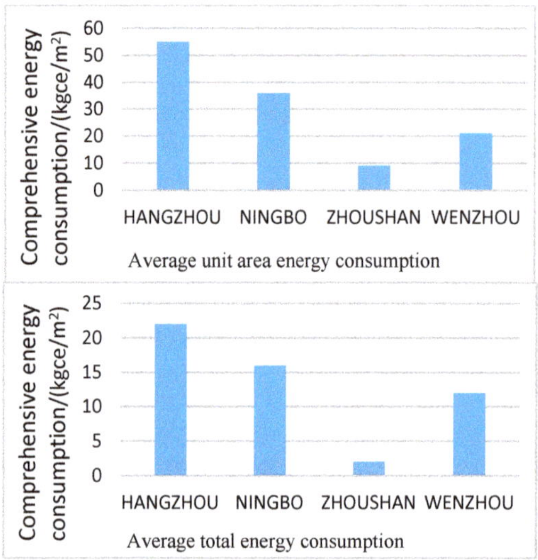

Figure 5. Comparison of the average energy consumption and average energy consumption of the average unit building area in each city.

3. Methods

Currently, the main method of suitability research is an indicator using the weight method, and currently, the method of determining the weight can be divided into the subjective weight method, the objective weight method, and the main objective weight method. The subjective weighting method is a kind of method in which researchers assign weights to each index according to their subjective value judgment. This kind of method is divided into the expert evaluation method, analytic hierarchy process, and so on. The weight of each index depends on the knowledge structure and personal preference of each expert. Although this is a good reflection of the subjective will, it lacks scientific stability. Considering its obvious defects, it is generally only applicable to the evaluation of data collection difficulties and information that cannot be accurately quantified. The objective weighting method is used to determine the weight according to the relationship between the original data, using a certain mathematical method. The judgment result does not depend on the subjective judgment of the person and has a strong mathematical theoretical basis [21]. Combination weighting method: Combining the results of subjective and objective weighting methods.

3.1. Method Selection

In the empowerment of the indicator system, the existing method of empowerment is both subjective and objective. Because this research is relatively pioneering, there are few domestic related researches. Therefore, the subjective empowerment method seems very weak. At this stage, the most widely used method of objective weighting is the entropy method [21–23].

Entropy was originally a thermodynamic concept. It was introduced by Claude Elwood Shannon to information theory, and it has been widely used in engineering, social, and economic fields [24–26]. In general, if the entropy of an index is smaller, the greater the degree of variation in the value of the target, the greater the amount of information, and the greater the role of the overall evaluation, the greater the amount of weight it has in the overall evaluation. On the contrary, the greater the entropy (E_j) of certain indicators shows the smaller the degree of variation of its index value, the smaller the amount of information provided, the smaller the comprehensive evaluation of the role, and its weight should be smaller. The information entropy calculation formula of the indicator j is as follows:

$$E_j = -(\ln m)^{-1} \sum_{j=1}^{m} p_{ij} \cdot \ln p_{ij} \tag{1}$$

where, m is the number of objects to be evaluated, and $p_{ij} = \dfrac{d_{ij}}{\sum\limits_{j=1}^{m} d_{ij}}$ if $p_{ij} = 0$, we define the following:

$$\lim_{p_{ij} \to 0} p_{ij} \cdot \ln p_{ij} = 0 \tag{2}$$

Scope of application: (1) can be used to determine the weight of the indicators in any evaluation question, and (2) can be used to eliminate the indicators that contribute little to the evaluation results in the indicator system [22]. Advantages: Objectivity, relative to those subjective valuation methods, the accuracy is higher and the objectivity is stronger, which can better explain the results obtained. Adaptability can be used for any process that needs to determine the weight, and can also be used in combination with some methods. Disadvantages: only the influence of the index variation on the weight is considered, and the conflict between the indexes is not considered [23]. To sum up, in order to make up for the shortcomings of the entropy method, we have chosen the CRITIC (Criteria Importance Though Intercrieria Correlation) method as the method of empowerment in this study. Its basic idea is to determine the objective weight of the indicators based on two basic concepts. First, the contrast intensity, which represents the value difference between the different evaluation schemes of the same index, is expressed in the form of standard deviation, that is, the standard deviation difference indicates the value difference between the different schemes within the same index. The larger the standard deviation is, the greater the value difference between different schemes will be. The second is the conflict between the evaluation indicators. The conflict between the indicators is based on the correlation between indicators. For example, there is a strong positive correlation between two indicators, indicating that the conflict between two indicators is low. The first j indicators and other indicators of the conflicting quantitative indicators are for $\sum_{i=1}^{n} (1 - r_{tj})$, r_{tj} is the correlation coefficient between evaluation index t and evaluation index j, and σ_j is the standard deviation of the index j. The determination of the objective weight of each index is a comprehensive measure of the comparative strength and conflict. If n is the number of evaluation indicators, let C_j denote the amount of information contained in the indicator j, then C_j can be expressed as follows:

$$C_j = \sigma_j \sum_{i=1}^{n} (1 - r_{tj}) \tag{3}$$

where, σ_j is the standard deviation of the jth index. The larger the C_j is, the more information the indicator j contains, and the greater the relative importance of the index is, so the objective weight W_j of indicator j should be as follows [27]:

$$W_j = \frac{C_j}{\sum\limits_{j=1}^{n} C_j} \tag{4}$$

3.2. Selection of Indicators

The evaluation index of passive energy utilization of public institutions is determined. When the selection of the index is taken into consideration, it considers both the supply side and the demand side and covers the characteristics of public institutions and the characteristics of energy [28]. According to the basic information and evaluation indexes of the passive energy utilization of public institutions, combined with the resource conditions in different climate zones and the characteristics of the energy use of typical public institutions, the research on the suitability of different types of passive energy is conducted so as to form the suitability distribution map.

Considering the characteristics of public institutions and passive energy, from the perspective of the supply-side, the energy distribution in various regions is extremely important. The amount of energy in a region can reflect how much energy the region can provide to public institutions, which plays a major role in the suitability research index system. From a macro point of view, the application of technology is not considered. Therefore, technical indicators related to passive energy, such as average sunshine duration and hydrogeological factors, are not considered in this study. In addition, economic factors should also be considered. This study selects regional fixed asset investment. Because public institutions are funded by the government to complete construction projects, the selection of fixed asset investment can reflect the size of the government's investment in the construction of the area.

From the demand-side perspective, it is necessary to understand the building's energy use characteristics. From the perspective of the nature of building energy consumption characteristics, it can be seen from the existing building climate zoning map that the energy consumption characteristics of all kinds of buildings are the same in each partition. Therefore, in the research on the passive energy suitability of public institutions, this research is carried out based on the climate zoning map. From the perspective of building energy consumption characteristics, the energy consumption index of public institutions is basically the same, so the number of public institutions can be used to reflect the size of the energy consumption.

The mathematical explanation for this is as follows: assume that the five climate zones are I1, I2, I3, I4, and I5. According to the characteristics of the climate zone, the energy consumption characteristics of the buildings in each climate zone are the same. The suitability of renewable energy in buildings is influenced by factors such as building energy consumption. The main component of building energy consumption is the building load. Estimating the building load of the entire city can be calculated by the area thermal index method. Assume that the number of cities studied in the I1 climate zone is m. As the cities are all in the I1 climate zone, the area heat index ($a_1 = a_2 = \ldots = a_m$) is the same, and the total area of the public institutions in the kth city is S_k. Therefore, the total load of all of the public institutions in the kth city is $F_k = a_k \times S_k$. In the later data processing, the most normalized method will be used, and the normalized data will be $F_k{}^*$, $F_k^* = \frac{F_{kmax}^* - F_k}{F_{kmax}^* - F_{kmin}^*} = \frac{a_k S_{kmax} - a_k S_k}{a_k S_{kmax} - a_k S_{kmin}} = \frac{S_{kmax} - S_k}{S_{kmax} - S_{kmin}}$. Assuming that the number of public institutions is n and the $\frac{S}{n} = \alpha$ of each province is the same, then $\frac{S_k}{n_k} = \alpha$ is the same. So, $F_k^* = \frac{S_{kmax} - S_k}{S_{kmax} - S_{kmin}} = \frac{\alpha n_{kmax} - \alpha n_k}{\alpha n_{kmax} - \alpha n_{kmin}} = \frac{n_{kmax} - n_k}{n_{kmax} - n_{kmin}}$. Thus, the main factor that ultimately affects energy consumption is the number of buildings.

In summary, this study conducted a suitability study on the basis of building a climate zoning map. The selected indexes were the average growth rate of the fixed asset investment, passive energy distribution in various regions, the number of public institutions, and energy consumption per unit area.

4. Research Process

In this study, the objective weighting method was used as the weighting method of the evaluation system, and the natural breakpoint method was used as the partitioning and data visualization method. Both of them are based on the characteristics of the data, excluding subjective factors to get the results, and they follow the scientific principle–the principle of objectivity, and the principle of feasibility by considering the characteristics of public institutions of passive energy suitability, in order to select the evaluation index. Then, the dirty data are removed by data features, and then the data are normalized. Secondly, the weight of the indicators is given, and the final score is obtained again. Finally, the natural breakpoint method was used to divide the country into 10 suitable areas (data visualization). In order to avoid subjective factors, CRITIC has made up the subjective disadvantages of the subjective empowerment law. This method has a strong objectivity and can indicate the degree of correlation (independence) between the indicators. At the same time, the natural breakpoint method is a statistical method for classification according to the objective data structure [29].

In the research on the passive energy suitability of public institutions, it is necessary to construct a corresponding evaluation system, which is generally required in order to give weight and score, and to find out the demarcation points for partitioning. Empowerment becomes the most important step. However, at the same time of empowerment, because of the great difference in the degree of data fluctuations, it is necessary to filter and eliminate the data, and then to conduct standardized processing. Therefore, the idea of this study is to screen and eliminate the data first, then carry out standardized processing, and finally carry out empowerment. In normalization, MMN (Min-max normalization) is a linear transformation of the original data, so that the resulting value maps are between (0–1) or some self-defined interval. The transformation function is as follows:

$$X^* = \frac{X - X_{\min}}{X_{\max} - X_{\min}} \tag{5}$$

$$X^* = \frac{X_{\max} - X}{X_{\max} - X_{\min}} \tag{6}$$

If the larger play a superior role in the index evaluation system, Equation (5) is used, otherwise Equation (6). Where, $X_{\max}$ is the maximum value of the sample data and $X_{\min}$ is the minimum value of the sample data. The disadvantage of this method is that when there are individual dirty data, such as an attribute value that is too large or too small, it may lead to changes in $X_{\max}$ and $X_{\min}$, which in turn affects the processing of the entire data. At this time, only the greatest value normalization can be redefined. The extreme value has the exclusion of dirty data [22].

In mathematics, the difference of two orders of magnitude can be regarded as the infinity of the higher order ratio and the infinitesimal of the lower order ratio. Therefore, it is necessary to consider the fluctuation degree of data when selecting indicators, and to try to select the one with the fluctuation degree within two orders of magnitude. If it is not within two orders of magnitude, data filtering and elimination are required. It can be observed that in an indicator, the interval within two orders of magnitude is taken to cover most data, and the interval is (a_i, b_i), where, $b_i/a_i < 100$. Let p_i be the data of the indicator. If $p_i > b_i$, the value takes 1 after normalization. If $p_i > a_i$, the value takes 0 after normalization. The calculation is shown in Tables 1–3.

Table 1. Some provinces' and cities' geothermal energy data processing.

Province	Geothermal Energy [MJ/(°C × m^2)]	Standardization (Unchecked Dirty Data)	Standardization (Checked Dirty Data)
Fujian	157.73	0.01	0.19
Gansu	160.09	0.01	0.19
Guangxi	506.46	0.02	0.88
Hainan	565.13	0.03	1.00
Jiangsu	413.38	0.02	0.70
Qinghai	63.62	0.00	0.00
Sichuan	454.55	0.02	0.78
Tianjin	496.89	0.02	0.86
Xinjiang	18764.20	1.00	1.00
Zhejiang	532.76	0.03	0.94

Note: The formula for the standardization of this indicator is Equation (5) [30]. Of a total of 31 sets of data, we selected 10 sets of data.

Table 2. The original data of each indicator of some provinces.

Province	Average Growth Rate of Fixed Asset Investment	Number of Public Institutions	Solar Energy Multi-Year Average (0.01 MJ/m^2)
Fujian	0.10	117,063	1.47
Gansu	−0.20	69,259	1.46
Guangxi	−0.40	86,813	1.68
Hainan	0.03	26,451	1.66
Jiangsu	0.11	127,608	1.38
Qinghai	0.14	41,668	1.61
Sichuan	0.16	103,631	1.03
Tianjin	0.19	15,678	1.57
Xinjiang	−0.02	54,331	1.64
Zhejiang	0.11	44,668	1.22

Table 3. The normalization of various indicators in some provinces and cities.

Province	Average Growth Rate of Fixed Asset Investment	Number of Public Institutions	Geothermal Energy	Solar Energy
Fujian	0.85	0.09	0.19	0.68
Gansu	0.34	0.52	0.19	0.66
Guangxi	0.00	0.36	0.88	1.00
Hainan	0.72	0.9	1.00	0.97
Jiangsu	0.86	0	0.70	0.54
Qinghai	0.91	0.77	0.00	0.89
Sichuan	0.95	0.21	0.78	0.00
Tianjin	1.00	1	0.86	0.83
Xinjiang	0.64	0.65	1.00	0.95
Zhejiang	0.85	0.74	0.94	0.30

Note: As the quantity of public organizations are reflected by the energy consumption, and the lower the quantity of energy consumption, the formula for standardization selection is Equation (6).

As the number of public institutions can objectively reflect the local energy consumption, the amount of energy can objectively reflect the local passive energy supply. Therefore, the weight of these two indicators is large, as shown in Figure 6.

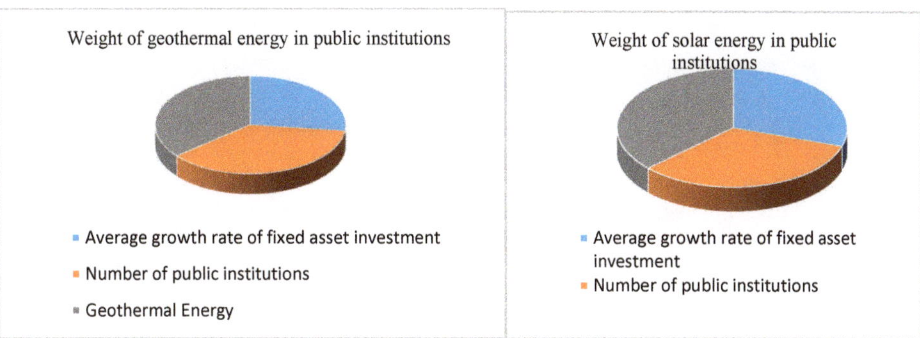

Figure 6. Weight diagram of passive energy for public institutions.

The final score for each region is as follows:

$$F = W \cdot X^*$$

(7)

The calculation results are as shown in Figure 7.

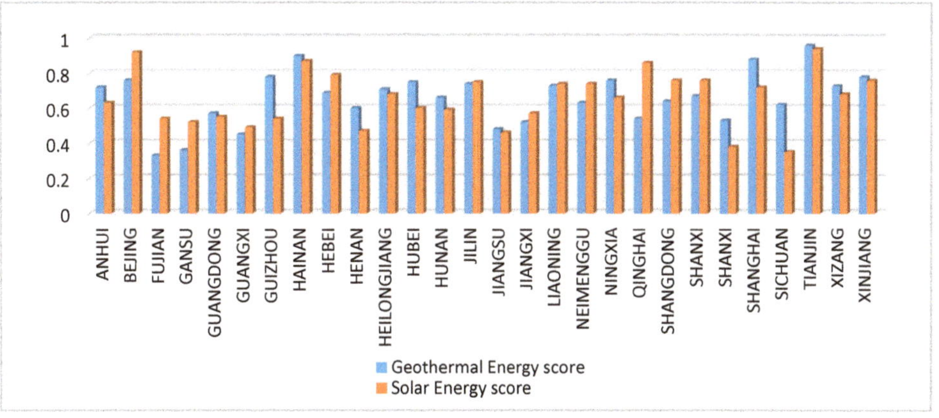

Figure 7. Public institution passive energy score map.

Finally, the data are partitioned and visualized in the form of a graph. The method used is the natural breakpoint method in ArcGIS. The natural breakpoint method is a statistical method that classifies according to the statistical distribution law, which can maximize the difference between classes. There are some natural turning points and feature points in any statistical series. With these points, the research objects can be divided into similar groups. Therefore, the crack itself is a good boundary for grading. Generating statistical data into frequency histograms, slope curves, and accumulated frequency histograms all help to find natural cracks in the data. This can be understood as follows:

1. The smallest difference within the class, the largest difference between classes
2. A map grading algorithm that considers that the data itself has a breakpoint and can be graded using the characteristics of the data.
3. The algorithm principle is a small cluster. The clustering end condition is that the variance between groups is the largest and the variance within the group is the smallest.

A suitability map was formed according to the above method.

5. Research Results and Analysis

The passive energy suitability of public institutions is based on the characteristics of public institutions to conduct zoning research. Therefore, it is necessary to classify public institutions with the same architectural characteristics into a class of suitability research. Therefore, on the basis of building climate zoning, suitability zoning was carried out. Each building climate region was divided into three sub-regions (suitability region, general suitability region, and unsuitable region), which serve as the passive energy suitability region of public institutions in the building climate region. Then, the suitable areas, general suitable areas, and unsuitable areas of different climate zones were classified into passive energy suitability divisions of national public institutions, and then the appropriate areas, general suitable areas, and unsuitable areas of the country were divided into 10 different areas. The Figures 8–11 show the following: (1) the suitability map of passive energy use in public institutions was roughly the same as the trend of the energy distribution maps, that is to say, places with more passive energy sources are more suitable for public institutions. (2) The suitability map was based on the architectural climate map as a base map, so it contains the energy consumption characteristics of the building. From the details, the trend of the suitability map of passive energy use in public institutions was not exactly the same as the trend of energy distribution maps. (3) The development of buildings to the upper levels has led to a significant decline in the utilization of solar energy in government office buildings, but the solar energy in Yunnan is abundant, so government office buildings are more suitable for solar energy [12]. This conclusion is consistent with the distribution map I have drawn. (4) Taking Beijing as an example, Guanghui Xu evaluated the geothermal energy in Beijing, and listed a large number of engineering cases to prove the suitability of ground source heat pumps in public institutions in Beijing [31]. This conclusion is consistent with the distribution map I have drawn. The distribution map is shown in the Figures 8–11. The higher the level of appropriateness, the more appropriate, that is to say, the darker the color, the more appropriate.

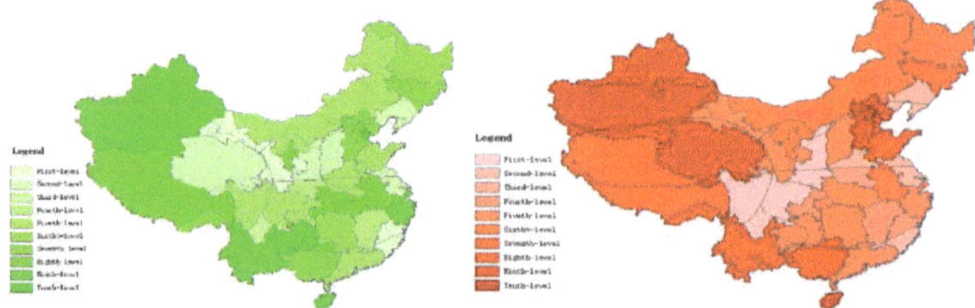

Figure 8. Distribution map of the geothermal energy (left) and solar energy (right) of the public institutions.

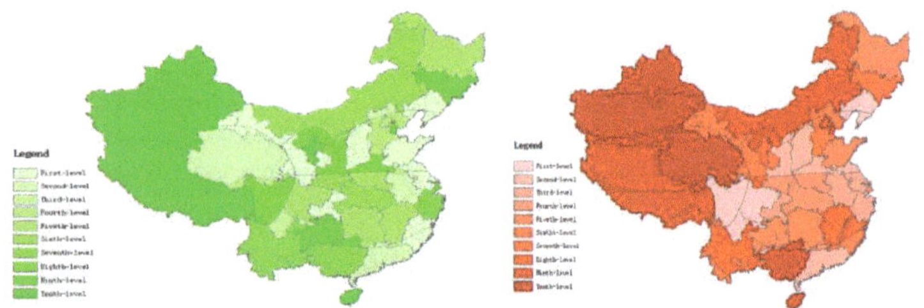

Figure 9. Distribution map of the geothermal energy (left) and solar energy (right) of schools.

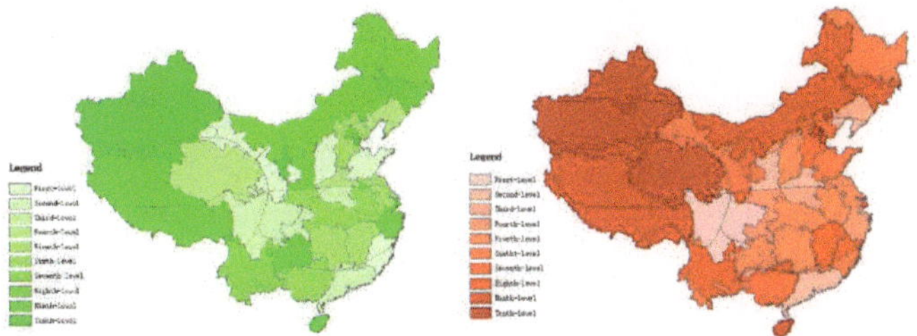

Figure 10. Distribution map of the geothermal energy (left) and solar energy (right) of government buildings.

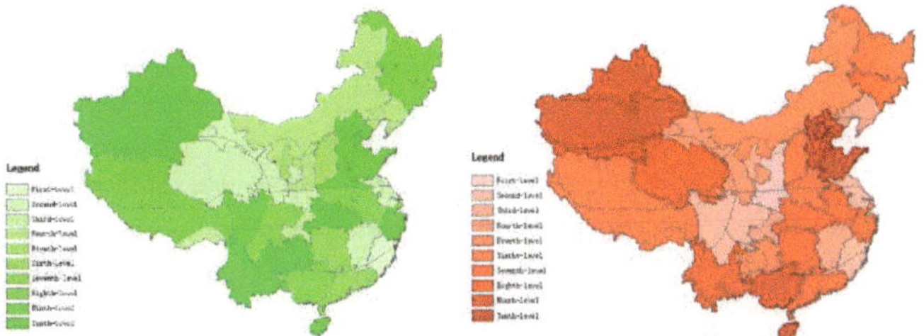

Figure 11. Distribution map of the geothermal energy (left) and solar energy (right) of hospitals.

Author Contributions: Conceptualization, S.Y. and H.L.; formal analysis, H.L.; funding acquisition, S.Y.; investigation, H.L., L.B., and F.H.; methodology, H.L.; project administration, S.Y.; resources, L.B.; writing (original draft), H.L.; writing (review and editing), H.L.

Funding: This research was funded by the National Key R&D Program of China, grant number 2017YFC0702601; the Shenyang Young and Middle-Aged Science and Technology Innovation Talents Program, grant number RC170313; and the Liaoning Provincial Natural Foundation, grant number 20170540761.

Conflicts of Interest: The authors declare no conflict of interest.

References

1. Xu, Q. Research Status and Thinking of Energy Consumption Quota for Public Buildings. *Build. Technol.* **2010**, *8*, 35–37.
2. Xue, S.; Li, Y.; Ma, Y.; Yu, Y. Research on energy consumption quotas in universities in Hubei Province. *Build. Sci.* **2013**, *29*, 93–97.
3. Zhang, W. Energy-saving design and operation strategy of air-conditioning cold and heat source in Shanghai Fengxian Central Hospital. *HVAC* **2013**, *43*, 52–54.
4. Tang, P.; Ling, S.; Yang, Y.; Cui, Q.; Gong, Y. Measurement and Analysis of Buried Tube Heat Exchangers in Nanjing Area. *Sol. Energy* **2017**, *38*, 378–385.
5. Li, J.; Du, S.; Lai, Z. Application Test Analysis of Soil Source Heat Pump Air Conditioning System in Guiyang Area. *Sichuan Build. Sci. Res.* **2015**, *41*, 204–207.
6. Sun, W.; Zhou, N.; Huang, J.; Cai, W.; Wang, Y. Analysis of the variation characteristics of the geothermal field in the heat transfer zone of the ground source heat pump system. *Acta Energ. Sin.* **2017**, *38*, 2804–2810.
7. Hu, W.; Yan, D.; Liu, W. Study on the Guide to Energy Consumption Limits of Wuhan Civil Buildings. *Build. Sci.* **2015**, *31*, 42–47.
8. Lai, M. *GB/T5116-2016, Energy Consumption Standard for Civil Buildings*; Ministry of Housing and Urban-Rural Development of the People's Republic of China, China Building Industry Press: Beijing, China, 2016.
9. 2018 BP World Energy Statistical Yearbook. Available online: http://www.bp.com/en_US/china/ (accessed on 23 May 2019).
10. Hou, E. Release of the annual development research report on building energy conservation in China 2018. *Build. Energy Conserv.* **2008**, *46*, 133.

11. Zhang, D.; Ding, Y.; Xie, L.; Kuang, Y. Investigation and analysis of energy consumption in government office buildings in chongqing. *Civ. Eng.* **2012**, *34*, 271–274.
12. Liu, Y.; Wu, D.; Li, H.; Lin, J.; Tang, L.; Cui, Y. Preliminary Analysis of Energy Consumption Statistics of Civil Buildings in Yunnan Province. *HVAC* **2016**, *46*, 6–11.
13. Long, W. Reflections on Building Energy Efficiency 2.0. *HVAC* **2016**, *46*, 1–12.
14. Yuan, X.; Gu, X.; Wang, J. Research Progress in Solar Energy Resource Assessment in China. *J. Guizhou Meteorol.* **2011**, *35*, 1–4.
15. Regulations on Energy Conservation in Public Institutions. Order of the State Council of the People's Republic of China. Available online: http://www.nea.gov.cn/2017-11/03/c_136725251.htm (accessed on 1 March 2019).
16. Chen, Y.; Wang, J.; Lin, W.; Fei, R.; Zhao, D. Study on the Suitability of Renewable Energy in Office Buildings in Zhejiang Province. *Zhejiang Archit.* **2016**, *33*, 50–54.
17. Libz, D. Part load operation coefficient of air-conditioning system of public building. *Energy Build.* **2010**, *42*, 1902–1907.
18. Liu, M. Research on Energy Consumption Prediction Model and Energy Conservation Management System of Public Buildings in Universities in Xi'an. Master's Thesis, Xi'an University of Architecture and Technology, Xi'an, China, 2017.
19. Guo, C. Research on Energy Saving Potential and Energy Saving Measures for Hospital Buildings. Master's Thesis, Hebei University of Engineering, Handan, China, 2018.
20. Lu, W.; Chang, Y.-X. Study on the Energy Consumption Survey of Hospitals in Zhejiang Province and Its Influencing Factors. *Build. Sci.* **2010**, *26*, 48–51, 107.
21. Baake, P.; Boom, A. Vertical Product Differentiation, Network Externalities, and Compatibility Decisions. *Int. J. Ind. Organ.* **2001**, *19*, 267–284. [CrossRef]
22. Zhang, S.; Zhang, M.; Chi, G. Science and Technology Evaluation Model and Empirical Study Based on Entropy Weight Method. *Chin. J. Manag.* **2010**, *7*, 34–42.
23. Wang, K.; Song, H. A Comparative Analysis of Three Objective Weighting Methods. *J. Technol. Econ. Manag.* **2003**, *6*, 48–49.
24. Wu, L.; Tan, Q.; Zhang, Y. Network connectivity entropy and its application on network connectivity reliability. *Physics A* **2013**, *392*, 5536–5541. [CrossRef]
25. Zou, Z.; Yun, Y.; Sun, J. Entropy method for determination of weight of evaluating indicators in fuzzy synthetic evaluation for water quality assessment. *J. Environ. Sci.* **2006**, *18*, 1020–1023. [CrossRef]
26. Liu, L.; Zhou, J.; An, X.; Zhang, Y.; Yang, L. Using fuzzy theory and information entropy for water quality assessment in Three Gorges region, China. *Expert Syst. Appl.* **2010**, *37*, 2517–2521. [CrossRef]
27. Diakoulaki, D.; Mavrolas, G.; Papayannakis, L. Determining objective weights in multiple criteria problems: The critic method. *Comput. Oper. Res* **1995**, *22*, 763–770. [CrossRef]
28. Long, W. Demand Side Energy Planning Adapts to Supply Side Structure Reform-Written in Front of "Urban Demand Side Energy Planning and Energy Microgrid Technology". *HVAC* **2016**, *46*, 135–138.
29. Wang, C. Evaluation of the Main Stream Resources in the Middle and Lower Reaches of the Yangtze River. Ph.D. Thesis, Graduate School of the Chinese Academy of Sciences (Nanjing Institute of Geography and Limnology), Nanjing, China, 2000.
30. Wang, G.; Zhang, W.; Liang, J.; Pei, W.; Liu, Z.; Wang, Y. Evaluation of geothermal resource potential in China. *Chin. J. Earth* **2017**, *38*, 449–450, 451–459, 584.
31. Xu, G. Demonstration Research on Shallow Geothermal Energy Resources Evaluation in Beijing. Ph.D. Thesis, China University of Geosciences (Beijing), Beijing, China, 2007.

Article

Optimal Municipal Energy System Design and Operation Using Cumulative Exergy Consumption Minimisation

Lukas Kriechbaum *  **and Thomas Kienberger**

Chair of Energy Network Technology, Montanuniversitaet Leoben, Franz-Josef Straße 18,
A-8700 Leoben, Austria; thomas.kienberger@unileoben.ac.at
* Correspondence: lukas.kriechbaum@unileoben.ac.at; Tel.: +43-3842-402-5408

Received: 25 November 2019; Accepted: 26 December 2019; Published: 1 January 2020

Abstract: In developed countries like Austria the renewable energy potential might outpace the demand. This requires primary energy efficiency measures as well as an energy system design that enables the integration of variable renewable energy sources. Municipal energy systems, which supply customers with heat and electricity, will play an important role in this task. The cumulative exergy consumption methodology considers resource consumption from the raw material to the final product. It includes the exergetic expenses for imported energy as well as for building the energy infrastructure. In this paper, we determine the exergy optimal energy system design of an exemplary municipal energy system by using cumulative exergy consumption minimisation. The results of a case study show that well a linked electricity and heat system using heat pumps, combined heat power plants and battery and thermal storages is necessary. This enables an efficient supply and also provides the necessary flexibilities for integrating variable renewable energy sources.

Keywords: energy systems optimisation; exergy analysis; cumulative-exergy consumption minimisation; multi-energy systems; energy-system design; municipal energy systems

---

1. Introduction

Recent studies for Austria showed that the available potential for renewable energy sources (RES) is smaller than the current demand [1,2]. To reach the goal of a fully climate neutral society, imports of RES from other countries or local efficiency measures are necessary. In this context, exergy is a useful concept for identifying efficiency potentials. Although energy is subject to the law of conservation and can never be created or destroyed, exergy is the maximum useful work that can be extracted from a form of energy. It is consumed when brought to equilibrium with its surroundings, therefore it is a potential which describes the ability to cause change. It is the motive force that determines the flow of energy and it constantly deteriorates on its way through the energy system [3]. In the literature, there exist a variety of tools and methods [4,5] to identify and reduce exergy destruction and exergy losses. Their main aim is to increase resource efficiency.

The various forms of energy have different exergy contents, e.g., electricity is pure exergy, whereas low temperature heat has a very low exergy content. Most of today's used (fossil) energy carriers have a high exergy content. Data for 2016 shows that 50.7 % of the final energy consumption in Austria is used for heat applications [6]. 66.1 % of the heat is used for low temperature applications like domestic heating, hot water or air conditioning. The rest is used for steam or in furnaces in high temperature industrial applications. In a number of energy strategies of highly developed countries, the focus is on decarbonising the electric power generation [7], even though in the OECD member countries only 22.2 % of the final energy consumption is electricity [8].

A comprehensive decarbonisation of the energy system requires efficiency measures and a replacement of fossil fuels by renewable electricity [9]. In multi-energy systems (MES) several sectors (e.g., electricity, heat and transport) and energy carriers (e.g., electricity, natural gas, biomass) are considered in an integrated approach [10]. The coupling of energy sectors and their infrastructures using suitable technologies (e.g., heat pumps, CHPs, etc.) enables the utilisation of synergies, and provides the necessary flexibility for integrating variable RES. MES can also relieve the strain on the energy transmission and distribution infrastructure [11,12].

Today's typical way of energy system optimisation often focus solely on the electricity system, mainly aiming at an optimum dispatch of power plants and storages. The well-developed electricity grid makes electricity an easy to transport good and establishes operational competition between the individual electricity producers. The present approach delivers optimum operational strategies for the whole system, but does not investigate its optimum design. For heat the situation is different, as heat supply is a local issue and individual buildings or small heat grids are usually supplied by one or few plants. The main decision regarding energy and exergy consumption of heat supply is made during the system design process, and the technology selection. Therefore the main two research questions which occur when designing an exergy efficient MES, where heat and electricity sectors are linked:

- What is the optimum system design? How can it provide the necessary flexibility options for the integration variable RES?
- How can this system be efficiently operated to always meet the demand?

This calls for a model which combines planning and operational aspects [13]. The cumulative exergy demand (CExC) methodology [14] includes both the above outlined points. Next to the exergy consumed during operation it also considers the exergy consumed to create the energy system's infrastructure.

In this paper, we present the application of the CExC-minimisation on municipal energy systems. First we present the current research and the relevant literature on exergy analysis and energy system optimisation in Section 2. This is followed by an introduction into the concept of exergy, the CExC methodology and the optimisation approach in Section 3. Section 4 presents the results from applying the methodology on a municipal energy system. We close this paper with a comprehensive discussion of the results in Section 5.

2. State of Research

The term exergy was first mentioned by Rant in 1953 when he described the "technical working capacity" [15]. Today, the concept of exergy is used in different kind of fields in environmental science and technology. In this section, the basics of exergy are first presented and then a literature overview of the current tools and methods of exergy analysis is given.

2.1. Fundamentals: Exergy, Exergy Destruction and Exergy Losses

The first law of thermodynamics describes the energy conservation. The second law indicates the irreversibility of natural processes. This means that in any real process, exergy is consumed and entropy is created. The second law also provides information regarding the convertibility of energy forms and the direction in which a process proceeds. For example, electricity or mechanical work theoretically can be fully converted into any other form of energy, whereas for instance for heat the convertibility depends on the temperature difference. In general, energy E consists of a useful part exergy B and the useless part anergy A.

$$E = B + A \tag{1}$$

Exergy is the useful work that can be theoretically extracted from a form of energy when it is brought to equilibrium with its ambient conditions. Anergy in contrary cannot conduct any work. A well-known example for anergy is heat at ambient temperature.

Four different forms contribute to the total exergy of a system: potential exergy B_{pot} (system height relative to the environment), kinetic exergy B_{kin} (system velocity relative to the environment), chemical exergy B_{ch} (deviation of the chemical composition to the environment) and physical exergy B_{ph} (deviation of pressure p and temperature ϑ from the environment p_0, ϑ_0) [5]. Potential and kinetic energy are pure exergy; the chemical exergy can be approximated by using the lower heating value [16]. The physical exergy, B_{ph}, of a mass, m, can be calculated by the enthalpy, h, and entropy, s, of a system with its p and ϑ compared to the ambient conditions (T_0, s_0).

$$B_{ph} = [(h - h_0) - (\vartheta_0 s - \vartheta_0 s_0)] \cdot m \tag{2}$$

Thermodynamic inefficiencies in an energy system are either caused by exergy destruction B_D or exergy losses B_L [17]. A well known example for exergy destruction is the production of hot water by burning natural gas. Irreversible thermodynamic transformations cause exergy destruction B_D by entropy generation s_{gen}. For a system with the mass m these irreversibilities can be described by the Guoy–Stodola theorem (Equation (3)). As exergy is always dependent on a reference state; it is usually described by the reference pressure p_0 and reference temperature ϑ_0. Commonly this reference is the "standard atmosphere".

$$B_D = \vartheta_0 s_{gen} \cdot m \tag{3}$$

Exergy losses B_L are caused by exergy transfers over the system boundaries. That might be work, heat or physical streams that cannot be further utilised. Examples are heat losses in a district heat network or flue gas exhaust streams from boilers.

2.2. Exergy Analysis: Tools and Methodologies

So far, for exergy analysis several tools and methodologies have been developed [4,5]. Examples are the the cumulative exergy consumption [14], the exergetic cost theory [18,19], thermoeconomics [20] or the extended exergy analysis [21]. They all share the same major goal to help to improve the system design, even though they have different system boundaries. As exergy is the potential to conduct work, it is especially suited as a common base in MES where different energy forms with different exergy contents are compared [16].

Cumulative exergy consumption (CExC) and the exergetic cost methodology extend exergy analysis of a single process beyond its boundaries to include all processes from natural resources to the final product. They both use a "fuel–product concept", where for any system a fuel exergy and a product exergy can be defined. Their exergetic definitions depend on the requirements of the task [18]. CExC analysis was introduced by Szargut in 1957 [22] and includes all exergetic expenses from raw materials to the final product [23]. The exergetic cost theory was introduced by Valero et al. [18] and is defined as "the sum of exergy contained in all resources entering the supply chain of the selected product or process" [5]. In this case, the term "cost" is the exergetic expenditure and has not monetary relation. Even though both methods use a different formalisation, their results are equivalent [5].

Both methodologies are applied to different fields. Szargut et al. [14] proposed the the CExC method to improve the "cumulative degree of perfection of chemical processes". Applications in energy conversion deal with an oxy-fuel combustion plant [24], or an organic Rankine cycle for waste heat power generation [25]. Valero et al. [26] applied the exergetic cost theory to the CGAM problem [27] and represented the productive structure explicitly, which allows optimisation at a local level. Lozano and Valero [28] performed an exergetic cost analysis on a steam boiler in a thermal generating station. In this study, the authors analysed variations in the exergetic costs of the total product and their causes in order to draw conclusions on the boiler's real performance. As seen in Misra et al. [29], the application of exergetic costs to a LiBr/H_2O vapour absorption refrigeration system enables an approximate optimum design configuration.

The CExC-method was also applied to analyse the resource consumption on a larger scale with different countries and societies [30], for example the United States [31] and China [32]. On a city scale, it was applied to compare energy scenarios in the smart city planning of Milan [33]. Waste heat [34] and low temperature district heat systems [35] were also investigated using exergy analyses. Krause et al. [16] carried out optimum power flow calculations for a MES to maximise its operational exergy efficiency. We did not find any recent literature where the CExC- or exergetic cost method was applied on municipal energy systems.

3. Methodology

In this work, we adapt the CExC-method to determine an exergy optimal design of a municipal MES. This requires modelling the energy system, and the assessment of energy as well as materials streams flowing in and out of the system. The optimal system design is reached when the CExC reaches a minimum. Therefore, we need a precisely formulated objective function and to model all the necessary constraints [21].

3.1. Cumulative Exergetic Consumption

CExC includes all exergetic losses and exergy destruction from raw materials or energy carriers to their final utilisation. Therefore, it quantifies the consumption of primary resources embodied in a product or service [23]. In an energy system, the exergy expenditures are stored in the energy imports, the materials necessary to build the infrastructure and the locally produced RES. RES and the imported energy carriers are converted to the desired form of energy to supply the load. Therefore, the services produced are both the load as well as the excess energy (Figure 1).

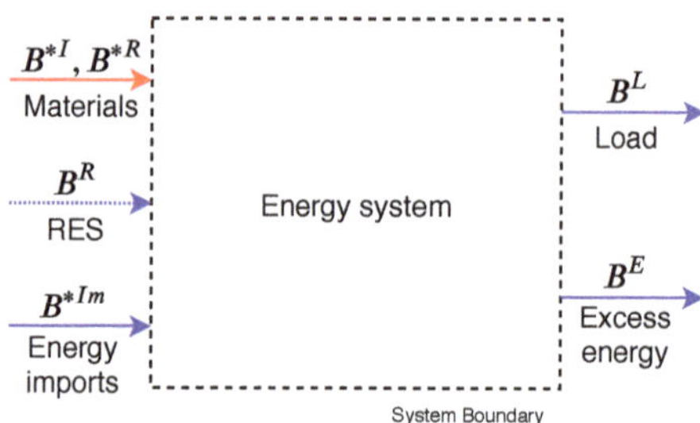

Figure 1. Material, RES, energy imports, load and excess energy flows over the system boundaries in an energy systems (source: own representation).

The imported energy flows may include renewable and nonrenewable sources. For the exergetic assessment of imported and exported energy following assumption are made [23]:

- Raw materials or energy carriers are attributed their reference exergy.
- Pre-treated materials or energy carriers are attributed an exergy content of their raw value and the exergetic expenses for the pretreatments.
- Any energy delivered to the load gets attributed its exergy content.
- Next to the energy produced to meet the load, excess energy might be created (Figure 1). If it can be further used, it is considered using its exergy content.

Therefore, for energy and material flows, we have to differentiate between their exergy content and their CExC. The exergy content B of an energy stream is the sum of its embodied physical, chemical, kinetic and potential exergy.

$$B = B_{pot} + B_{kin} + B_{ch} + B_{ph} \tag{4}$$

The "cumulative exergetic consumption" (CExC), B^*, of a stream is its exergy content, B, and all the exergy destruction, B_D; exergy losses, B_L; and exergetic expenses, B_{exp}, that occurred throughout the steps p of the production process PP of a stream. Any expenses are caused by irreversibilities within the production processes [19]. These can occur "directly or indirectly, in the extraction, preparation, transportation, pretreatment and manufacturing process" [21].

$$B^* = B + \sum_{p \in PP} \left(B_{D,p} + B_{L,p} + B_{exp,p} \right) \tag{5}$$

This means that the production route has an impact on the CExC of a product. The same product delivered by two different production processes can have a different CExC.

3.1.1. Unit Expenditures and Unit Content

In energy systems modelling, the descriptive variables are usually energy flows and the capacities of the installed infrastructure. Therefore, we use unit expenditures or unit contents to convert energy, E; materials, m; or capacities, C, to CExC or exergy. The unit expenditures describe the CExC per unit of energy, material or capacity. The conversion factor is called CExC-factor. The unit content describes the actual amount of exergy stored in a unit of energy. This conversion factor is therefore called exergy factor. The use of those conversion factors makes the CExC methodology applicable together with a broad range of energy system modelling tools.

The exergy factor r is used to convert energy to exergy. It is the proportion of exergy B in energy E:

$$r = \frac{B}{E} \tag{6}$$

The CExC-factor r^* is used to convert energy to CExC and describes the CExC per unit energy. It is the ratio between CExC B^* and energy E (Equation (7)). For materials the calculation is equivalent, but relative to a unit of mass m.

$$r^* = \frac{B^*}{E} \tag{7}$$

CExC for the different energy carriers can be taken from literature. In the cases where no data is available, the use of cumulative energy demand (CED)-values is also acceptable. This is applicable, because we only consider energetic resources in this paper. In such cases, CED and CExC-values have a comparable magnitude with a coefficient of determination of more than 99 % [36]. Therefore we assume CED and CExC-values to be identical.

CExC B_I^* of the infrastructure units is incorporated in the materials necessary to build these conversion units, RES and storages. Therefore the infrastructure-CExC-factor r^* can be defined as the CExC per capacity unit C of a conversion unit, a RES, or a storage:

$$r^* = \frac{B^*}{C} \tag{8}$$

For some energy technologies, the CExC-factors are directly available in literature. As the lifetime of infrastructure units is usually longer than the investigated period in the model, CExC are only taken into account proportionally. Therefore, we can define the equivalent periodic CExC-factor r^{*p}. It expresses the CExC B^* over the investigated period per unit installed capacity C.

$$r^{*p} = r^* \cdot \frac{T}{T_{LT}} \tag{9}$$

If CExC-factors are not directly available in literature they can be calculated using data from existing infrastructure units. CExC over the lifetime T_{LT} of an infrastructure unit with the nominal capacity C_n can be calculated based on its material consumption m_m, and the respective material CExC-factors r_m^*. We assume linear relations between the capacity and the required materials, as well as between the investigation period T and the lifetime T_{LT}. Thus, the equivalent periodic CExC-factor expresses the CExC for one unit of capacity for a certain period of time:

$$r^{*p} = \frac{\sum_m r_m^* m_m}{C_n} \cdot \frac{T}{T_{LT}} \tag{10}$$

3.1.2. Example CExC for Domestic Heat from an Electric Resistance Heater

For district heating, we want to produce heat with an exergetic content of 0.2 (physical exergy calculated using Equation (2) and based on following assumptions; feed temperature 70 °C and reference temperature 0 °C). The installed electric heater shall have a capacity of 1 MW and an efficiency of 99 %. In an investigation period of one year, the annually produced heat shall be equivalent to 2000 full load operating hours.

Electrical heater equivalent periodical CExC-factor: An electric heater with a nominal capacity of 10 kW and a lifetime of 15 a consists of 40 kg of steel. Steel has a CExC-factor of $1.75 \times 10^{-5} \frac{TJ}{kg}$. Material consumption of this boiler results in a CExC of 0.2 MW h. Therefore, using Equation (10) the equivalent periodic CExC for an electric heater is 1.3 $\frac{MW\,h}{MW\,a}$.

Electricity and heat CExC-factors and exergy factors: Even though the exergy content of electricity is 1, in Austria its CExC is 2.96. The annual heat production has an energy content of 1980 MW h and an exergy content of 396 MW h. The total exergetic expenditures (electricity and materials for the electric heater) add up to 5933.3 MW h per year. The exergy expenditures can be divided into 1.3 MW h for plant investment, 3920 MW h for pretreatment of the electricity and 2000 MW h for the electricity itself. Accordingly, the produced energy has a CExC-factor of 2.99, but only an exergy factor of 0.2.

For a plant with a given nominal capacity, the unit expenditures vary dependent on the annual full load operational hours. They include expenditures for the resistance heater and expenditures for the electricity. The latter ones consist of the expenditures, due to the consumption of physical exergy and the expenditures for the pretreatment. Pretreatment expenditures are those expenditures necessary to provide an energy carrier with its embodied physical exergy at the system boundaries. In this example, we assume that the pretreatment for electricity is constant. Also, the efficiency of the resistance heater is constant, which leads to a constant consumption of electricity per unit heat. Therefore the expenditures for physical exergy and pretreatment are independent of the full load operational hours and stay constant in Figure 2. For the expenditures from infrastructure investment it is different, because they only occur once during the investigation period. The more heat is produced with the resistance heater, the smaller becomes their share on the total unit expenditures (Figure 2). The cumulative unit expenditures for the electric heater with one full load operational hour are 4.3. For 8760 full load operational hours they decrease to 2.99. Then the share caused by plant investment is negligible.

3.2. CExC Minimisation of Multi-Energy Systems

The main objective in this paper is to design an energy system, which has minimum cumulative exergy destruction and exergy losses. We use a greenfield design approach, that means we do not consider existing energy infrastructure. A municipal energy system shall be designed for a given electricity and heat demand. On the one hand, exergy is needed in the form of materials to set up the physical infrastructure of the energy system. On the other hand, it is consumed in form of conventional energy carriers or RES to operate the system and supply the demand (Figure 3). In the case of high RES production, excess energy might be produced. The optimum system is reached, when the difference between exergy flowing into the system and exergy flowing out of the system gets a minimum.

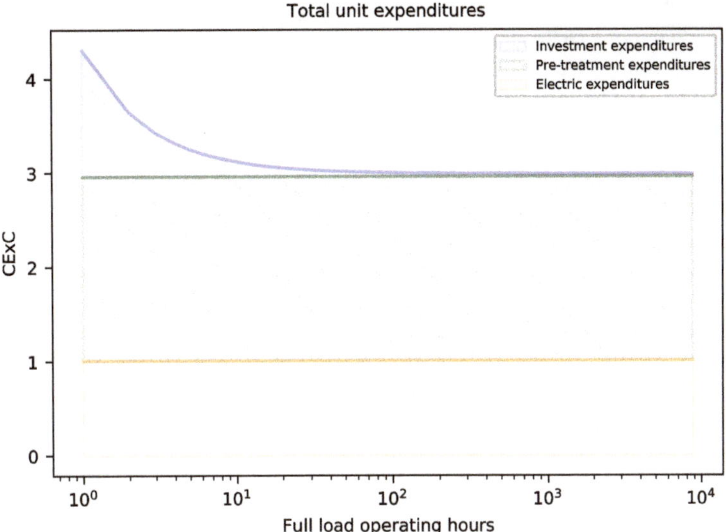

Figure 2. Total unit expenditures for investment, pretreatment and exergy consumption. (Source: own representation).

A model must be set up which includes all relevant boundary conditions, but still leaves a certain degree of freedom for optimisation. For this task, we use the optimisation framework *oemof*, which is specifically designed for energy system optimisation. The model must allow several different supply routes using RES, conversion units (e.g., power plants, boilers, grids, etc.) and storages (e.g., batteries, hot water tanks, etc.). For these infrastructure elements, the used materials and their CExCs must be specified. The operational boundary conditions include the imported energy (e.g., electricity, natural gas, biomass, etc.) and RES potentials (e.g., wind, PV) and their CExC.

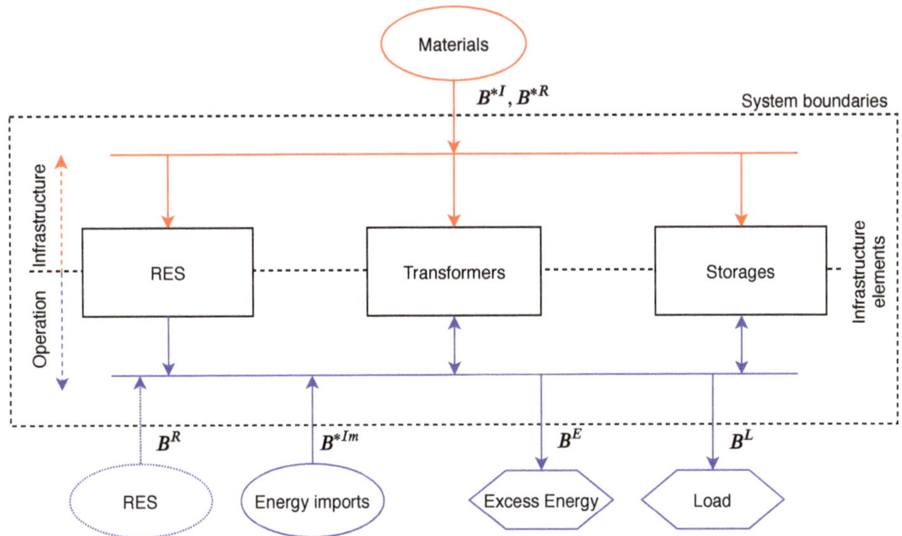

Figure 3. Cumulative exergetic consumption (CExC) flows, equivalent periodic CExC flows and exergy flows in an energy system. (Source: own representation).

3.2.1. *oemof*—Open Energy Modelling Framework

The Open Energy Modelling Framework (*oemof*) [37,38] is an open source framework for cross-sectoral and multi-regional modelling and optimising of MES. It can deal with multiple energy carriers, for example electricity, heat, biomass, natural gas, etc. In *oemof*, an energy system is represented by a graph, consisting of a set of edges and nodes. The edges are the logic links between the nodes, they describe the structure of the energy system. The nodes represent the technical components of the infrastructure. The components include all the main technical equipment of an energy system: sources, sinks, conversion units, grid elements and storages. The individual components can be connected via busses to each other.

Sources represent any imported energy, for example fuels, RES, natural gas and electricity from the grid. Sinks are used to model energy flows out of the energy system, for example, loads and electricity export. Energy conversion processes are described by conversion units, e.g., in power plants, boilers, etc. They can have multiple inputs (P^{in}) and multiple outputs (P^{out}) for a set of different energy carriers $\alpha, \beta, \ldots \in \Gamma = \{electricity, natural\, gas, biomass, heat, \ldots\}$ and are described by their conversion efficiencies η [16].

$$\begin{pmatrix} P_\alpha^{out} \\ P_\beta^{out} \\ \vdots \\ P_\omega^{out} \end{pmatrix} = \begin{pmatrix} \eta_{\alpha,\alpha} & \eta_{\beta,\alpha} & \cdots & \eta_{\omega,\alpha} \\ \eta_{\alpha,\beta} & \eta_{\beta,\beta} & \cdots & \eta_{\omega,\beta} \\ \vdots & \vdots & \ddots & \vdots \\ \eta_{\alpha,\omega} & \eta_{\beta,\omega} & \cdots & \eta_{\omega,\omega} \end{pmatrix} \cdot \begin{pmatrix} P_\alpha^{in} \\ P_\beta^{in} \\ \vdots \\ P_\omega^{in} \end{pmatrix} \tag{11}$$

Energy storage is modelled using a differential energy balance between the state of energies SOE of two consecutive time steps (Equation (12)). It includes inflow and outflow conversion losses (η^{in}, η^{out}) and standby losses η^{loss} over a time step τ.

$$\Delta SOE = \eta^{in} \cdot P^{in} - \eta^{loss} \cdot SOE^{t-1} \cdot \tau - \eta^{out} \cdot P^{out} \tag{12}$$

Transmission and distribution infrastructure (e.g., power lines, district heat or natural gas pipelines) are modelled like conversion units with a conversion efficiency. They have the same input and output energy carrier. For any component a nominal value, minimum and maximum values as well as an actual value including a time series can be defined.

3.2.2. Objective Function

Cumulative exergy losses and exergy destruction become a minimum when the difference between expenditures and yields are a minimum. The expenditures include all the consumed exergy: for RES B^{*R}, for imported energy B^{*Im} and for the infrastructure B^{*I}. The yields include all the useful produces exergy: the load B_L, and any excess exergy B_E due to the variable production of RES.

As exergy is a potential compared to a reference state, this reference state must be selected carefully taking into account the objectives of the model. Therefore, we define the following assessment guidelines for the exergetic assessment of any inflow and outflow streams.

- All flows into the energy system get attributed their CExC-factor.
- All flows out of the energy system get attributed their exergy factor.
- Any form of energy becomes valuable as soon as it has a common usable and transportable form, therefore when it is secondary energy (e.g., electricity, district heat, natural gas, hydrogen or biomass). We do not assign the raw energy forms like solar irradiation or the kinetic energy stored in the wind speed any exergy. This is consistent with the international recommendations for energy statistics [39]

Based on these guidelines and the system boundaries specified in Figure 3, the objective function for the energy system over the time period T can be formulated as follows.

$$min(B^{*Im} + B^{*I} + B^{*R} - (B^L + B^E)) \tag{13}$$

In *oemof*, the descriptive variables of streams and infrastructure units are power flows P and capacities C. With the help of time step τ power is converted to energy. CExC-factors and exergy factors described in Section 3.1.1 are used to convert these variables to exergies. All flows need to be summed up over the investigation period $T = \{t_{start}, \dots t_{end}\}$.

All the imported energy flows must be converted to the CExC B^{*Im} for the period T:

$$B^{*Im} = \sum_{t \in T} \sum_{i \in Im} P_i^S(t) \cdot r_i^{*S}(t) \cdot \tau \tag{14}$$

Assessment of the outflows is analogous, but this time we use the exergetic value instead. The consideration of the outflows is only necessary, because we consider the excess energy E as valuable.

$$B^L = \sum_{t \in T} \sum_{j \in L} P_j^L(t) \cdot r_j^L(t) \cdot \tau \tag{15}$$

$$B^E = \sum_{t \in T} \sum_{k \in E} P_k^E(t) \cdot r_k^E(t) \cdot \tau \tag{16}$$

CExC calculation for all infrastructure elements I is based on the capacity C, and the equivalent periodic CExC $r^{*,p}$.

$$B^{*I} = \sum_{l \in I} r_l^{*p} \cdot C_l \tag{17}$$

According to guideline (3), RES R are different. Electricity P^R from RES is seen as an exergy expenditure and rated with its exergy factor. Operational upstream exergy losses, for example, from solar irradiation to electricity, are not taken into account. Next to the operational exergy expenditures, the infrastructure investment must also be considered. CExC is treated as analogous to the other infrastructure investments. Consideration of operational expenditures is necessary to avoid CExC-factors for electricity from RES, which are lower than its respective exergy factors.

$$B^{*R} = \sum_{t \in T} \sum_{m \in R} P_m^R(t) \cdot r_m^R(t) \cdot \tau + \sum_{m \in R} r_m^{*p} \cdot C_m \tag{18}$$

3.3. Result Evaluation

The major results are the installed capacities of conversion units, storages, RES, the energy consumption from the grids and the excess energy produced. We also calculate total CExC for energy inflows and the exergy outflows. To rate the operational performance of conversion units, the capacity factor c_l of any conversion unit l can be calculated:

$$c_l = \frac{\int_{t=0}^{t_{end}} P_l(t)}{P_{l,inst} t_{end}} = \frac{E_l^{out}}{E_l^{out,max}} \tag{19}$$

It compares the energy E_l^{out} a conversion unit produces during a certain period to the maximum energy $E_l^{out,max}$ a conversion unit could produce during this time.

The storage cycles c_s for any storage s show how often an energy storage is fully charged or discharged. It is the discharged energy E_s^{out} during a certain period divided by the installed storage capacity $C_{s,inst}$

$$c_s = \frac{\int_{t=0}^{t_{end}} P_s^{out}}{C_{s,inst}} = \frac{E_s^{out}}{C_{s,inst}} \tag{20}$$

4. Case Study

For our case study, we use the presented methodology in a greenfield approach to determine the optimum design of a municipal energy system. A greenfield approach means to model the energy system from the scratch. No existing infrastructure is considered. Energy loads, RES characteristics, exergetic indices and an available set of energy conversion technologies and storages are given. To account for the different shares of RES in the electricity from the grid, four different scenarios with different CExC-indices are evaluated. For any scenario a model is created in *oemof* and the results are discussed.

4.1. System Description

The medium-sized model city is located in a region attractive for wind power and PV installations, but has no potentials for run of the river hydro power or pumped hydro. Our case study focuses on municipal energy systems, therefore it considers electricity, process heat and domestic heat demand from the residential, commerce and public services sector. Industrial demand is not encompassed in our case study, because such consumers are mostly supplied by transmission grids and not by municipal distribution grids.

The energy system is connected to the electrical and natural gas transmission grids. RES potentials, biomass potentials and waste heat from an industrial process are available. For the sake of simplicity, we use an aggregated representation of the municipal energy system. All the individual conversion units, energy storages, RES, energy sources and energy loads of one kind are lumped together to a single one. This aggregation process is carried out according to the "cellular approach" [40]. To account for distribution grid losses, energy production and domestic consumption are modelled in two different regions or so called cells (Figure 4). Both are connected by electrical power lines and district heat networks.

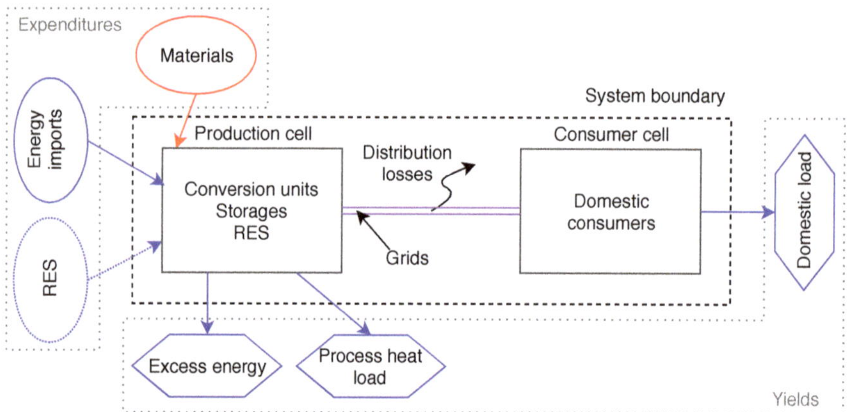

Figure 4. Two cell model with the exergy expenditure and yields. Conversion units, storages, renewable energy sources (RES) and the process heat load are located in the production cell. All domestic consumers are located in the production cell. (Source: own representation).

The values to be determined are the nominal capacities from conversion units, storages, RES as well as the imported energy and the excess electricity produced. Input parameters for the model are the loads, the available technologies, the maximum RES potentials and the possible conversion routes. In addition, CExC-factors and exergy factors must be provided for all specified technologies and energy carriers.

The electricity grid connection is bidirectional, that means that energy can be imported and exported. Even though in reality this is a single unit, in *oemof*, it is modelled using a source for imports and a sink for the excess electricity with a maximum connection power (Tables 1 and 2). Natural gas, waste heat and biomass are unidirectional, and therefore modelled using a source (Table 1). Natural gas and waste heat also do have a maximum capacity. The local biomass potential equals 22.5 GW h and must be fully exploited. Because biomass has no energy transport restriction like grid based energy carriers and can be easily stored, no maximum capacity is prescribed. CExC-factors for electricity, natural gas and biomass are taken from literature Table 1. Because there was no value available for waste heat, we estimated the CExC-factor based on its physical exergy using Equation (2) (assumptions: feed temperature 70 °C and reference temperature 0 °C).

Table 1. Energy sources: grid connection capacities and CExC-factors.

Imported Energy	Max. Capacity MW	CExC-Factor $\frac{MW}{MW}$
Electricity	60	$r^*_{el} = 2.96$ [41]
Natural gas	60	$r^*_{g} = 1.12$ [42]
Waste heat	3	$r^*_{wh} = 0.21$
Biomass		$r^*_{b} = 1.10$ [42]

Table 2. Excess energy: grid connection capacities and exergy factors.

Excess Energy	Max. Capacity MW	Exergy-Factor $\frac{MW}{MW}$
Electricity	60	$r_{el} = 1.00$

In our case study, we look at the domestic sector as well as at the businesses and commercial services sector. The annual demands specified in Table 3 include electricity, process heat, as well as domestic heating and hot water. Domestic heat has a temperature of 70 °C. Process heat is consumed by businesses and commercial services for the production of goods. The mean application temperature is assumed with 273 °C. Using Equation (2), this leads to the exergy factors specified in Table 3.

Table 3. Loads: annual demand, maximum power and exergy factors.

Load	Annual Demand GW h	Max. Power MW	Exergy Factor $\frac{MW}{MW}$
Electricity	55	11.9	$r_{el} = 1.00$
Domestic heat	90	38.2	$r_{dh} = 0.20$
Process heat	9	2.2	$r_{ph} = 0.50$

In *oemof*, all loads are modelled as sinks with fixed time series. We used the load profile generator *oemof demandlib* [43] to create load profiles with a resolution of 15 min from annual demand values . The required temperature data was retrieved from *renewables.ninja* [44,45] which uses the MERRA-2 data set. Exemplary for our model data of the year 2014 and the location of Eisenstadt (city in the eastern part of Austria; latitude: 47.84°, longitude: 16.54°) will be used. *oemof demandlib* uses standardised BDEW load profiles for modelling domestic and process heat time series [46,47] and electricity time series [48]. We assume that 30 % of the heat is used in single family houses, 40 % in multi family houses and the remaining 30 % in small businesses, commerce and services. The domestic heat demand is calculated for windy conditions and includes the hot water consumption. 20 % of the electricity is consumed by households, the remaining 80 % by small businesses, commerce and services. The electric load profiles do not include any demand for hot water production, as this is already considered in the domestic heat load profiles.

All conversion units, storages, RES and the process heat load are located in the production cell. Energy conversion is modelled using Equation (11), energy storage using Equation (12). All the available energy conversion, energy storage and RES technologies in the model are listed in Tables 4–6. Note that the biomass boiler, gas boiler and resistance heater are made available for both the production of domestic heat and process heat. In addition, we assume that 20 % of the high temperature process heat are waste heat, and can be further used for domestic heating. The specification parameters are conversion efficiencies, charge and discharge efficiencies, standby losses, maximum RES capacities, and equivalent periodic CExC per unit installed capacity. A detailed derivation (including all the references) of the equivalent periodic CExC for the individual units can be found in Appendix A. Normalised time series for PV and wind yields are retrieved from *renewables.ninja* using the same location and year as for the demand. Grid losses are modelled using transmission efficiencies and the networks do not have a restricting capacity (Table 7)

Table 4. Conversion units: considered conversion technologies.

Technology	Efficiency	Equivalent Periodic CExC-Factor $\frac{\text{MW h}}{\text{MW a}}$
Biomass boiler [49]	$\eta_{th} = 0.85$	$r^{*p}_{th} = 8.14$
Gas boiler [50]	$\eta_{th} = 0.95$	$r^{*p}_{th} = 6.83$
Heat pump [51]	$COP = 3$	$r^{*p}_{th} = 2.60$
PEM eletrolyser [52]	$\eta_{H_2} = 0.8$	$r^{*p}_{el} = 126.68$
PEM fuel cell [52]	$\eta_{el} = 0.8$	$r^{*p}_{H_2} = 126.68$
Resistance heater [53]	$\eta_{th} = 0.99$	$r^{*p}_{th} = 1.30$
Biomass CHP [54]	$\eta_{th} = 0.5\ \eta_{el} = 0.35$	$r^{*p}_{el} = 81.5$
Gas CHP [55]	$\eta_{th} = 0.5\ \eta_{el} = 0.35$	$r^{*p}_{el} = 24.34$

Table 5. Considered energy storage technologies.

Technology	Inflow Efficiency	Outflow Efficiency	Capacity Loss	Equivalent Periodic CExC-Factor $\frac{\text{MW h}}{\text{MW h a}}$
Battery storage [56]	$\eta^{in} = 0.86$	$\eta^{out} = 0.86$	$\eta^{l} = 10^{-8}$	$r^{*p}_{el} = 16.42$
Thermal energy storage [57]	$\eta^{in} = 0.99$	$\eta^{out} = 0.99$	$\eta^{l} = 2 * 10^{-4}$	$r^{*p}_{dh} = 4.19 \cdot 10^{-1}$
Hydrogen storage [58]	$\eta^{in} = 0.98$	$\eta^{out} = 0.98$	$\eta^{l} = 10^{-8}$	$r^{*p}_{H_2} = 1.24$

Table 6. Considered variable RES.

Technology	Max. Potential MW	CExC-Factor $\frac{\text{MW}}{\text{MW}}$	Equivalent Periodic CExC-factor $\frac{\text{MW h}}{\text{MW a}}$
PV [59]	60	$r_{el} = 1$	$r^{*p}_{el} = 347.6$
Wind [60]	33	$r_{el} = 1$	$r^{*p}_{el} = 67.1$

Table 7. Transmission efficiencies of the energy grids.

Grid	Efficiency
Electricity	$\eta_{el} = 0.99$
Domestic heat	$\eta_{th} = 0.85$

Using all the specified data, the objective function is composed according to Equation (13).

4.1.1. Sensitivity Analysis with Respect to the Electricity Source CExC

Electricity from RES has a lower CExC-factor compared to today's prevalent thermal generation. This is because RES do not include the exergy destruction expensive conversion from chemical to thermal energy. Also the assessment guidelines (see Section 3.2.2, guideline three) support this, as the the produced electricity are the exergy expenditures and not the raw energy form like wind or solar irradiation. Therefore, the proceeding integration of RES into the future electric energy system will lead to decreasing CExC-factors for electricity from the grid. As these are relevant design parameters for the model, the different scenarios will lead to different optimum system designs.

An accurate value for future CExC-values cannot be determined at the present. Therefore, we will carry out calculations for four different scenarios, starting with the reference case SR. It describes the current state for the CExC-factor for electricity in Austria [41]. The following scenarios S1, S2 and S3 represent future electricity systems with higher shares of RES (Table 8). The other parameters stay the same in all four scenarios.

Table 8. Electricity CExC-factor for the sensitivity analysis.

	SR	**S3**	**S2**	**S1**
CExC in $\frac{MW}{MW}$	2.96	2.00	1.50	1.25

4.2. Results

The high exergy expenditures for imported electricity in the reference case SR lead to the highest total exergy expenditures (Figure 5). They also make investments into conversion units, storages and RES worthwhile. This leads to the higher expenditures for investment and RES as well as fewer energy imports. The large installed capacities of variable, non-dispatchable RES also generate more excess electricity.

At times when the grid connection is not a limiting factor, the CExC-factor for electricity from the grid determines the maximum unit expenditures for local electricity generation. The unit expenditures are influenced by the CExC-factor of the used energy carrier, the investment expenditures, the efficiency and the capacity factor (compare to Figure 2). Only technologies which comply with this limit will be selected, otherwise the energy will be drawn from the grid. Therefore, the lower CExC-factors in S3, S2 and S1 will not allow for an infrastructure investment as extensive as in SR. This leads to reduced total exergy expenditures and a shift from infrastructure investment to energy imports. Due to the lower installed RES capacities, excess electricity also decreases in those scenarios.

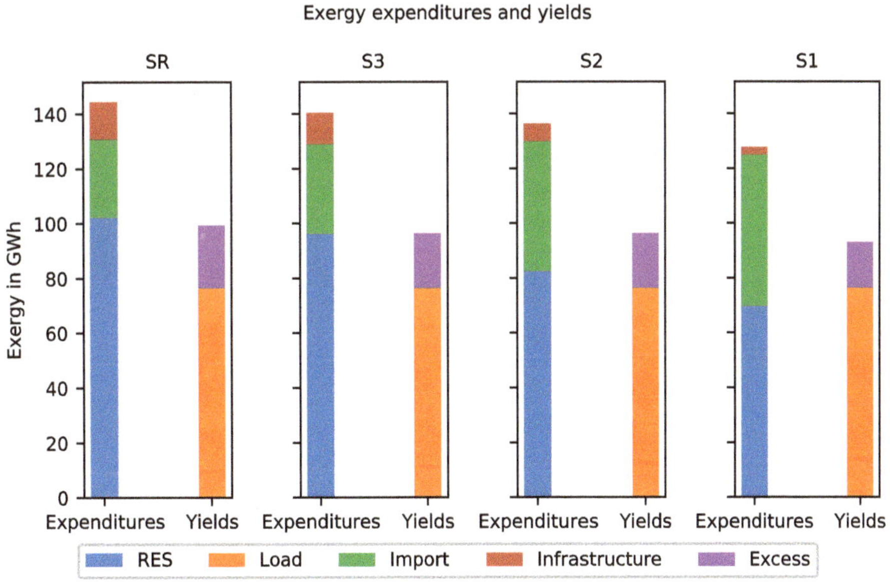

Figure 5. Exergy expenditures and yields of the different scenarios. Expenditures include CExC from energy sources, RES, and for RES and infrastructure investment. The yields include the exergy for the load and excess exergy. (Source: own representation.)

The following sections provide further details on installed capacities and operation of conversion units, RES and storages for all four scenarios. Afterwards the operational exergy expenditures and exergy yields are presented, followed by a discussion of the results.

4.2.1. Infrastructure Capacities and Expenditures

The capacities and the corresponding CExC from installed conversion units and RES are shown in Figure 6. Available technologies described in Section 4.1 that have not been selected for deployment, are not shown in the results. All the displayed capacities relate to the power produced (e.g., heat for the heat pump and boilers, electricity for RES). In the case of the CHP, which produces heat and electricity, the nominal electrical output is displayed.

Compared to the other conversion units, the high installed capacities of heat pumps and wind are apparent. PV and CHP capacities rise with an increase in the CExC-factors in the scenarios. While wind power is expanded to its maximum potential in all scenarios, PV never uses its maximum potential. A PEM electrolyser and fuel cell are installed only in SR. Biomass boilers and gas boilers are only used to supply the process heat load, but not for domestic heat. Even though RES do not have the highest installed capacities, their CExC exceeds the expenditures for conversion units by several orders of magnitude in all scenarios.

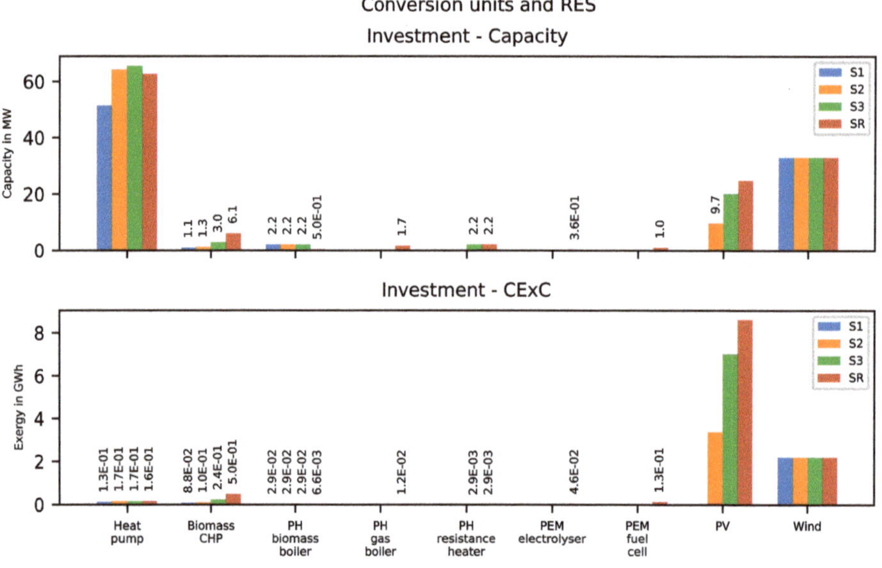

Figure 6. Calculated optimum conversion unit capacities and the corresponding CExC (Source: own representation).

All the different conversion units and storages are operated exergy efficiently and depending on the overall composition of the system. Even though we used 15-minute mean values in our model, we use daily mean values to present the results for unit operation in Figure 7. This provides a better visualisation of the long-term results. In this case, for the period of a whole year.

The heat pump provides domestic heat all year long except for the summer month. The biomass CHP operates mainly during times with a high heat demand and a low PV yield. At the same time, process heat in S3 is produced by a biomass boiler, and in SR, it is produced by a biomass and gas boiler. In the complementary times, the process heat is provided by a biomass or gas boiler and a resistance heater, which is operated with excess electricity from PV or wind (Figure 7). In S1 and

S2, high temperature heat is provided by biomass boilers and resistance heaters. The electrolyser is predominately operated in the second half of summer and in autumn. The conversion back to electricity takes place at the beginning of the year and in the second half of the year.

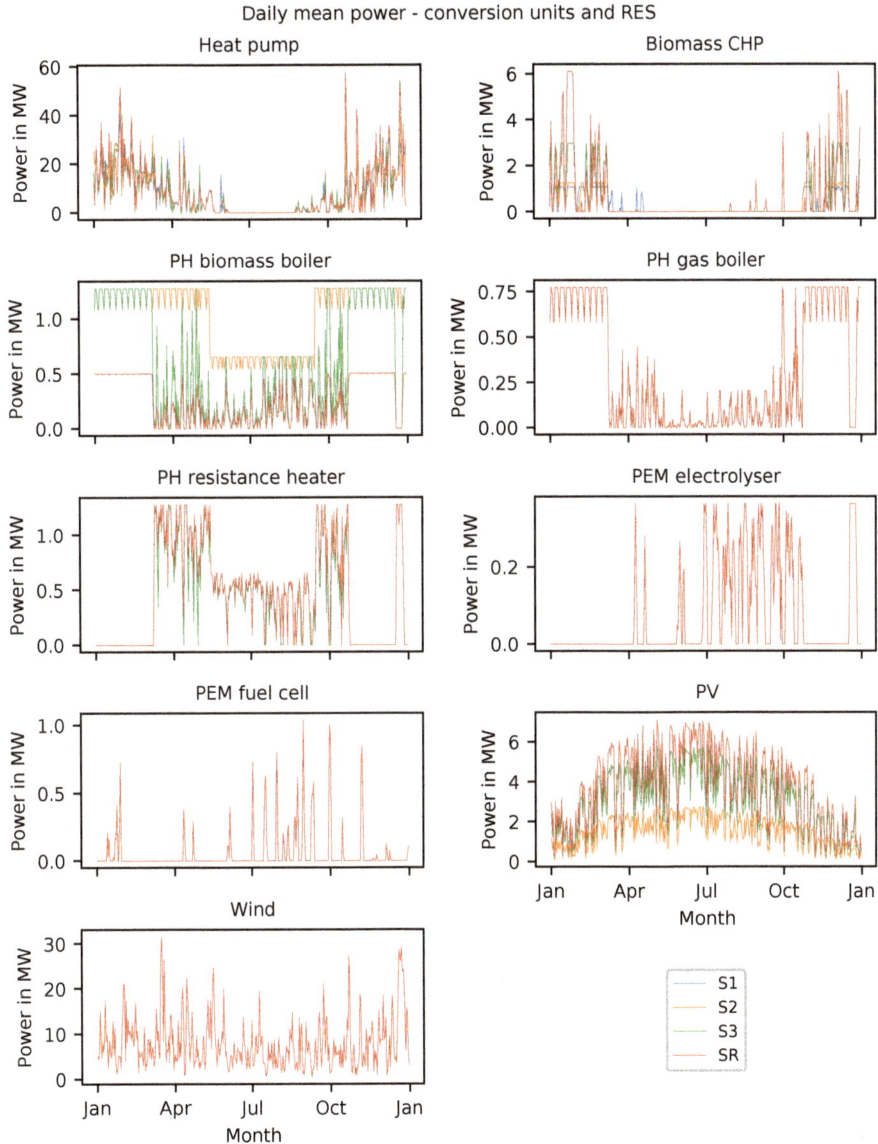

Figure 7. Daily mean power of the conversion units and RES. (Source: own representation.)

The more different conversion units available, the lower the capacity factors (Table 9), which are calculated according to Equation (19). Exceptions are small scale units with dedicated base-load operation, for example, the process heat biomass boiler in SR. Because of the major seasonal component of domestic heat and hot water demand, the capacity factors for the production units are restricted by the shape of the load profile. The same applies for the electrolyser and the fuel cell. They are part of the long term H_2 storage and only one can operate at a time.

Table 9. Capacity factors of conversion units and RES.

	S1	S2	S3	SR
Heat Pump	16.6	13.4	12.9	13.2
Biomass CHP	27.5	23.5	15.9	10.0
PH biomass boiler	46.4	46.4	27.0	51.4
PH gas boiler	-	-	-	18.1
PH resistance heater	-	-	19.4	20.7
PEM electrolyser	-	-	-	20.0
PEM fuel cell	-	-	-	5.2
Wind	24.1	24.1	24.1	24.1
PV	14.9	14.9	14.9	14.9

The installed storage capacities are shown in Figure 8. The thermal storage capacity is several orders of magnitude larger than the battery and the hydrogen storage. Even though, the CExC for batteries and thermal energy storages are of a comparable magnitude. Hydrogen storage only makes exergetically sense in scenario SR. Remarkable is the vast increase of battery and thermal energy storage increase between the scenarios S2 and S3.

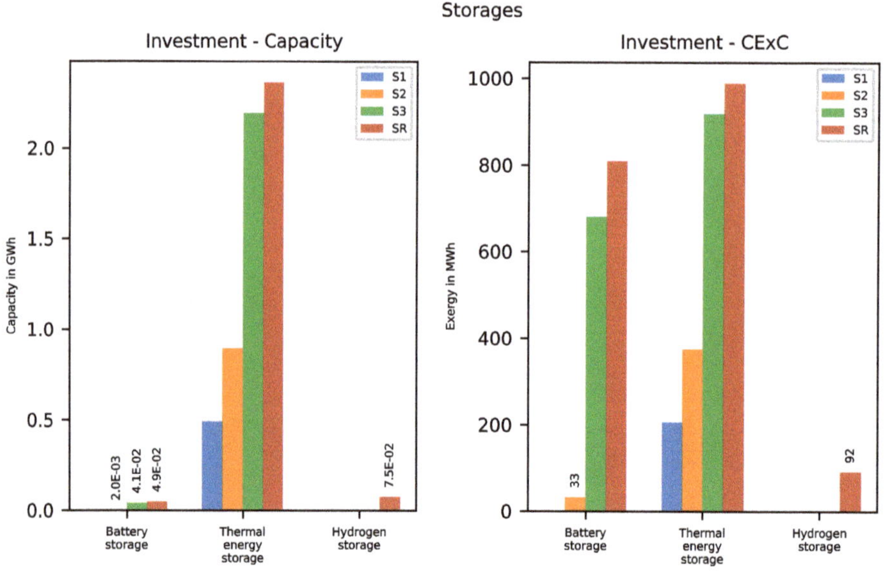

Figure 8. Calculated optimum storage capacities and the corresponding CExC. (Source: own representation).

The storage facilities are operated exergy-efficiently to bridge the gap between variable RES production and demand. Figure 9 shows the daily mean state of energy. With the help of the discrete Fourier transform (DFT), the state of energies time series can be decomposed into their individual periodical components. The results in Figure 10 show the amplitude and the numbers of cycles per year. Components which are smaller than 15 % of the maximum amplitude are removed from the plots.

In all four scenarios, the battery shows the highest states of energy during spring and autumn. During summer, the storage cycles are shorter, but the mean states of energy are also lower (Figure 9).

The DFT analysis shows clearly defined annual (one cycle per year) and daily cycles (365 cycles per year) in all three scenarios in which a battery is installed. The thermal energy storage is mainly used during the heating season with similar peak states of energy in the beginning and the end of the year. Exceptions are S3 and SR, where the peaks in autumn are more than twice as high compared to the spring. In all four scenarios, the amplitude of annual cycle is clearly dominant. The amplitude of this annual cycle is all the more significant with larger installed storage capacities. The hydrogen storage starts to get charged in July to shift electricity from the sunny periods to autumn and winter. Its state of energy has significant annual and biannual cycles.

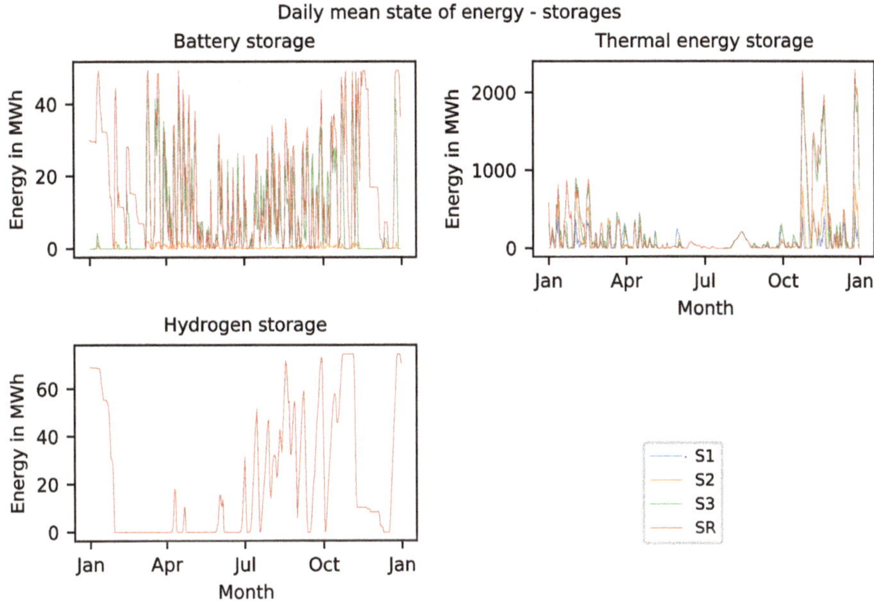

Figure 9. Daily mean states of energy. (Source: own representation).

The full cycles that a storage can achieve depends on load and production time series, as well as the size and purpose of a storage (Table 10). The battery in S2 has more than twice as many full cycles compared to the ones in S3 and SR. The thermal storage has the most utilisation during the heating season and is barely utilised in summer. Although the thermal storage peak states of energies are in the same order of magnitude for spring and autumn in S1 and S2, they are more than twice as high in autumn for scenarios S3 and SR. Excess electricity is used by heat pumps to shift the excess energy from PV over longer periods to times with higher demand and less supply. This requires higher storage capacities where large shares of the total capacity are not very often used. This and the great demand difference between summer and winter leads to significantly less storage cycles compared to the battery. Even though the hydrogen storage is a seasonal storage technology, it has 8.4 full storage cycles per year.

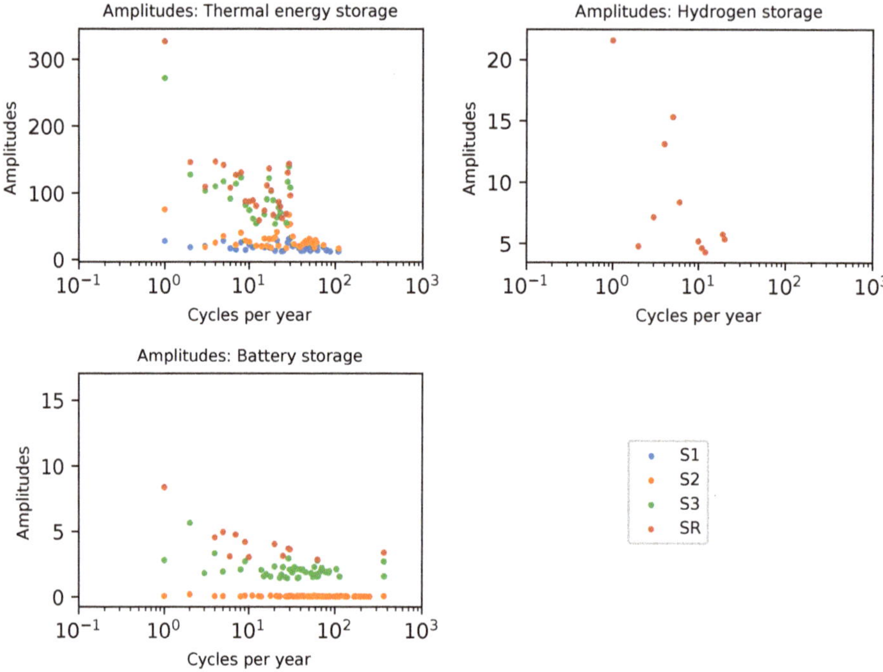

Figure 10. Periodical components of the state of energy. (Source: own representation.)

Table 10. Full storage cycles per year.

	S1	**S2**	**S3**	**SR**
Battery storage	-	140.1	68.6	67.2
Thermal energy storage	37.7	22.2	11.5	10.3
Hydrogen storage	-	-	-	8.4

4.2.2. Energy Imports, Excess Energy and Loads

Figure 11 shows annual energy and exergy loads, excesses, imports and the RES production. Loads, waste heat imports, biomass imports and electricity production from wind stay constant in all four scenarios. Imported electricity and natural gas, PV production and excess energy vary across scenarios with the CExC-factor for imported electricity. The higher the CExC-factor, the higher are PV-production and excess electricity, and the lower are the electricity imports. In scenario SR where the CExC-factor is the highest, no electricity is imported, but natural gas. Also, it is clearly visible that the exergy content of domestic heat is low when comparing annual energy and exergy loads of the domestic heat.

Figure 12 shows the daily mean power for loads, electricity excess, and energy imports. Electricity and process heat loads mainly fluctuate over days and weeks, the annual variations are secondary. For the domestic heat load it is different. Its major annual fluctuation is caused by its strong temperature dependency. Biomass and gas imports are the highest when the heat load is highest as well. The waste heat is consumed to its maximum extent, except for short periods in summer.

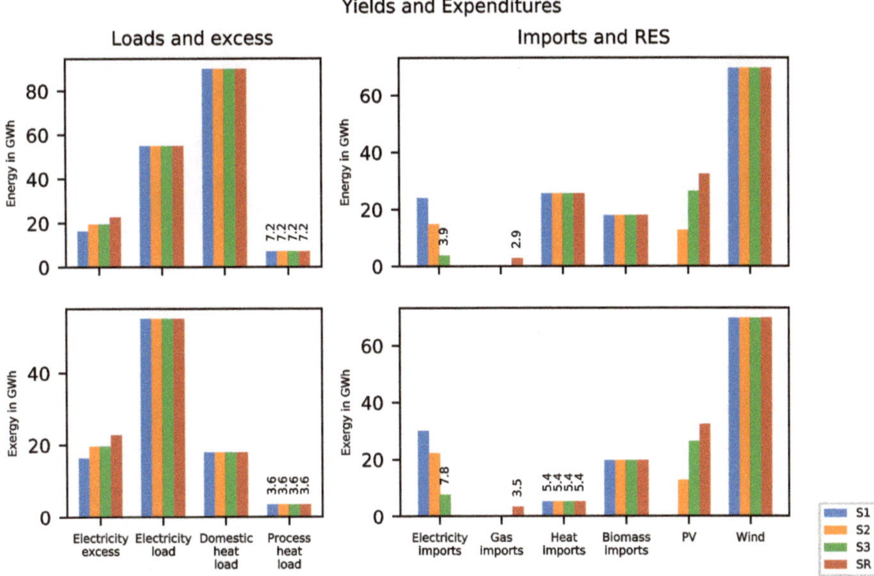

Figure 11. Energy sources and energy load, and the corresponding CExC, exergy load and excess exergy. (Source: own representation.)

Daily average values for electricity imported from the grid show a high variability. Although there are days with very little to no consumption, those days can be followed by peaks up to an average of 12 MW. The highest daily average values for electricity drawn from the grid occur during the winter months. In the summer months, those peaks drop to half of those values for S1 (Figure 12). This spread increases with increasing electricity CExC-factor until no electricity is consumed in SR. Excess electricity is produced between March and November, and in winter in case of high wind production. For SR Figure 12 shows that the exported electricity decreases from its peak in spring until autumn.

4.2.3. Result Discussion

Compared to imported energy, RES can provide energy for lower expenditures, but their fluctuating production does not necessarily meet the current demand. This gap must be compensated by energy drawn from the grid, or additional local power plants and storage facilities. Between the two options, the choice depends on unit expenditures for energy imports, and the investment as well as operational expenditures for conversion units and storage facilities. The unit expenditures of the imported energy carriers limit the maximum unit expenditures of local energy production, as long as there is no import capacity restriction. The local unit expenditures for an energy carrier include expenditures for conversion units and storages, and for exergy destruction and losses. The results reflect this context in higher total expenditures and a shift from operating to infrastructure expenditures in scenarios with higher CExC factors for imported electricity. Therefore, of all the scenarios, SR has the highest installed capacities of RES and conversion units (Figure 6), and is the only one where a long-term hydrogen storage makes sense (Figure 8).

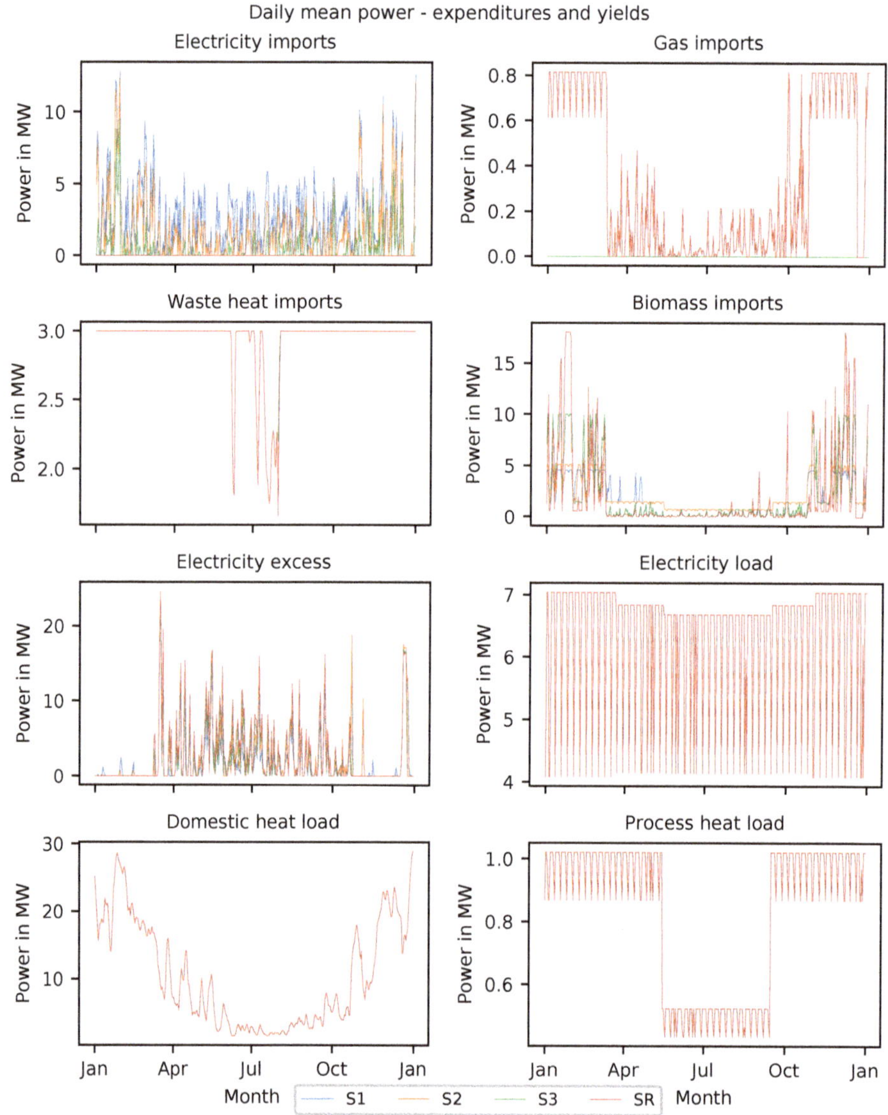

Figure 12. Daily mean power of the operational expenditures and yields. (Source: own representation.)

In our model, electricity imports can be seen as unrestricted, because the maximum load is well below the maximum grid capacity (Tables 1 and 3) This means that local production is only preferred if it has lower expenditures than the energy imports. In the case of excess electricity from RES, it can be stored locally for later use or it can be returned to the grid. For a useful storage investment, unit expenditures for electricity from RES and the battery must be lower than for imported electricity. The yield for electricity export must be also considered. This is the context that leads to the installed capacities of RES and storages. In all four scenarios, the wind power potential is used to its maximum. No PV is used in S1, but it rises up to 24.8 GW in SR, which is equal to 99.1 % of the available potential.

The higher CExC-factors for imported electricity make PV installations and battery storage practical in S2, S3 and SR. Long term storage using power to gas is exergetically only reasonable in SR.

For domestic heat, the situation is different. The maximum waste heat import power of 3 MW covers only 8 % of the maximum domestic heat load. The remaining heat will be provided by the plants with the lowest total unit expenditures, under the consideration that the local biomass has to be used. The biomass is used in a biomass CHP which is mainly operated in times where the heat demand is high and PV yields are low. Heat pumps together with thermal energy storage cover the rest.

From S2 to S3 the CExC-factor and therefore the unit expenditures for electricity imports rises from $1.5 \frac{MW}{MW}$ to $2 \frac{MW}{MW}$. This results in an increase of the battery storage capacity from 2 MW h to 39.5 MW h and of the thermal energy storage from 895 MW h to 2193 MW h. Apart from the two scenarios, there are no others where the increase in storage capacity is so large. As already discussed above, raising the limiting unit expenditures for electricity imports allows for higher total unit expenditures. The increase can be totally attributed to the infrastructure expenditures, because the operational unit expenditures stay constant. This tolerates lower capacity factors or annual storage cycles. Due to the fact that the investment unit expenditures follow a reciprocal function (see Figure 2), such a vast increase of the storage capacities between these two scenarios is possible.

5. Conclusions

The method of the CExC minimisation has proved to be applicable to energy systems planning and operation. The usage of CExC-factors makes the methodology applicable to a wide range of modelling tools. Aggregation concepts like the "cellular approach" [40] allow for the deployment on different spatial scales with different levels of accuracy. Even though we presented a greenfield design approach, it is also well suited for brownfield design approaches, for unit commitment, and even for optimal power flow calculations.

The major point of the overall results is that a well linked electricity and heat sector, using heat pumps and thermal energy storage, can enable a resource efficient supply while providing the necessary flexibility for integrating variable RES at the same time. Co-generation of heat and electricity is beneficial to separate production. The second point is the consideration of the load collective of the different plants. Although the operational efficiency of one technology might be higher compared to another, its high investment CExC makes this technology more costly in cases with low annual capacity factors.

In general, for the same rated power an exergy efficient plant will be larger compared to a less exergy efficient one (e.g., compare the sizes of compression and absorption chillers.) [61]. This is because exergy efficient plants have to operate with lower driving potentials, which leads to larger plants and therefore to higher CExC for the plant investment. For example, for the same heat transfer capacity, heat exchangers with higher temperature differences between the hot and the cold fluid need smaller exchange surfaces than heat exchangers with lower temperature differences. This means that operational expenditures shift to investment expenditures. Therefore, exergy efficient plants need higher capacity factors than less exergy efficient plants to reach the same unit expenditures. The variability of RES requires additional storages and dispatchable back-up plants with high capacities. This will lead to low annual capacity factors for the individual plants, which contradicts the use of exergy efficient technologies. In such cases, investment expenditures might not be negligible any more compared to the operational expenses. The CExC methodology takes both discrepancies into account and supports finding an optimal solution.

Although exergy factors for energy streams can be unambiguously calculated by thermodynamic laws, we know that the CExC-factors for the inflows do not have such a high degree of accuracy and are subject to uncertainties. The influence of the investment CExC should also not be overestimated, because in none of the scenarios it exceeds 10 % (Table 11). For a comparison, the conversion losses (exergy losses and destruction within the energy system) range from 21.8 % to 25.1 %. A sensitivity analysis of the investment CExC of the individual plant can help to get a better understanding of their implications.

Table 11. Share of investment and conversion losses on the total CExC input.

	S1 %	S2 %	S3 %	SR %
Investment	2.1	4.6	8.0	9.4
Conversion losses	25.1	24.7	23.4	21.8

Most of the current applications of technical exergy analyses differ in two points: the assessment of consumed energy carriers and materials according their physical exergy or CExC, and the use of either an energy-based or power-based perspective. Energy-based means that only the energy consumption over a certain period of time (usually one year) is considered for analysis. A power-based perspective also takes into account the variation in energy consumption over time [11].

Both of the above-stated points can have significant impacts on the relevance of the results. The use of physical exergy as an evaluation criterion could favour energy sources with high exergy losses outside the system boundaries over those produced internally. A power-based approach, as we use it in our work, is important for the sizing of system components and, in the case of involved storages, for considering their operational impact. Applications using physical exergy and a power-based approach mainly concern individual industrial processes or plants to identify internal exergy destruction and losses [5]. Energy-based perspectives are used above all when larger energy systems, in which the individual processes are no longer comprehensible, are considered over a longer period of time. Some consider only the flows of energy carriers, others include both energy and material flows [30].

Municipal energy systems lie between these two extremes. Our approach with using CExC-minimisation for design and operation of such energy systems helps to overcome this gap. It contributes to the identification of the location and magnitude of high exergy destruction and losses within the system. However, the use of CExC shows whether these are better treated within or outside the system boundaries. Including the materials for the plant investment permits optimum sizing of the plants for the respective load collective.

In future modelling applications, there are several improvements that can be made:

- A better representation of the **electric grid connection**. In this paper, we use a constant CExC-factor for electricity from the grid. We assume grid availability for feed in and drawing energy all the time. This might not be true for future applications. The CExC-factor might vary over the day and the seasons. There might also be shortages or congestion in the transmission grid.
- The **spatial dimension**. The current model does not account for the distribution of energy. Restricted energy transport capacities and the unavailability of network coverage in some areas, especially heat and gas grids, will lead to different results. The network restrictions can be modelled using total transfer capacities, or if more detail is necessary, power flow models.
- Include further **technologies**. Currently, only a basic set of conversion technologies (Table 4), RES (Table 6), and storages (Table 5) is used in the model. Possible additional technologies are demand side management; absorption heat pumps; solid oxide fuel; and electrolysis cells, pumped hydro, tidal energy, etc. The use of storage capacities inherited in heat and gas networks can also support the integration of RES.
- Include additional **sectors**. Currently, only the residential sector, and commercials, private and public services sector are considered in the model. Together they consume 34.3 % of Austria's final energy demand. The other large consumers are the industry sector with 29.3 % and the transport sector with 34.4 %. An incorporation of both sectors into a municipal energy system model can support in finding an exergy efficient design. A better model of industrial processes

can lead to synergies between industrial and municipal energy systems. In addition, a shift to electric mobility will increase electricity demand and include a high DSM potential.

Author Contributions: The authors contributed as following to the manuscript: conceptualisation, L.K. and T.K.; methodology, L.K.; software, L.K.; resources, L.K.; writing—original draft preparation, L.K.; writing—review and editing, T.K.; visualisation, L.K.; supervision, T.K.; project administration, T.K.; funding acquisition, T.K. All authors have read and agreed to the published version of the manuscript.

Funding: This research received no external funding.

Acknowledgments: I want to thank my colleague Philipp for all the support creating the plots.

Conflicts of Interest: The authors declare no conflicts of interest.

Abbreviations

The following abbreviations are used in this manuscript:

CED	cumulative energy demand
CExC	cumulative exergy consumption
CHP	combined heat and power
DFT	discrete Fourier transform
OECD	Organisation for Economic Co-operation and Development
RES	renewable energy sources
A	anergy
B	exergy
B^*	CExC
c_c	capacity factor
c_s	annual full storage cycles
E	energy
h	enthalpy
η	efficiency
m	mass
P	power
r	exergy factor
r^*	CExC-factor
r^{*p}	equivalent periodical CExC-factor
s	entropy
t	time series
T	time period
τ	time step
ϑ	temperature

Appendix A

Equivalent Periodic CExC

Equivalent periodic CExC-factors for infrastructure units are calculated using Equations (9) or (10). All data regarding RES, conversion units and storages is available in Tables A1–A5. Material CExC-factors are composed of exergy demand for RES, non-RES and other energy sources and can be found in Table A6.

Table A1. CExC-factors and lifetime for PEM fuel cell and electrolyser.

	CExC-Factor $\frac{\text{MW h}}{\text{MW}}$	Lifetime a
PEM fuel cell [52]	1900.2	15
PEM eletrolyser [52]	1900.2	15

[a] Because PEM electrolysis and fuel cell technology are comparable, we assume the same CExC-factors for both.

Table A2. CExC-factor and lifetime for the Li-ion battery.

	CExC-factor $\frac{\text{MW h}}{\text{MW h}}$	Lifetime a
Battery storage [56]	328.3	20

Table A3. Capacity, lifetime and material data for conversion units—part I.

	Capacity kW	Lifetime a	Steel kg	Concrete kg	Organic PVC kg	HDPE kg
Gas boiler [50]	10	15	200		10	
Gas CHP [55]	250	15	5000	50,000		
Wind [60]	500	20	50,000	300,000		7500
Biomass CHP [54]	800	20	48,000	800,000		

Table A4. Capacity, lifetime and material data for conversion units—part II.

	Capacity kW	Lifetime a	Steel kg	Copper kg	Silicon kg	Aluminium kg
PV [59]	300	30	60,000	1500	30,000	5400
Resistance heater [53]	10	15	40			
Heat pump [51]	10	15	80			
Biomass boiler [49]	50	15	1250			

Table A5. Capacity, lifetime and material data for storages.

	Capacity kW h	Lifetime a	Steel kg	Concrete kg	Organic PVC kg	HDPE kg
Thermal energy storage [57]	466	50	970		230	
Hydrogen storage [58]	1	50	13			

Table A6. CExC-factors of the used materials of the considered conversion units and storages.

Material	CExC-other TJ/kg	CExC-Renewable TJ/kg	CExC-non Renewable TJ/kg	CExC TJ/kg
Steel [62]	2.66×10^{-6}	1.15×10^{-8}	1.49×10^{-5}	1.75×10^{-5}
Organic PVC [63]	-5.7×10^{-7}	4.6×10^{-6}	1.36×10^{-5}	1.76×10^{-5}
Concrete [64]	5.18×10^{-8}	1.42×10^{-7}	4.62×10^{-6}	4.81×10^{-6}
HDPE [65]	2.82×10^{-7}	1.03×10^{-7}	1.18×10^{-5}	1.22×10^{-5}
Aluminium [66]	3.24×10^{-6}	3.16×10^{-5}	1.05×10^{-4}	1.40×10^{-4}
Copper [67]	1.80×10^{-6}	3.70×10^{-6}	3.38×10^{-5}	3.93×10^{-5}
Silicon [68]	7.63×10^{-6}	5.05×10^{-5}	2.55×10^{-4}	3.13×10^{-4}

The results for conversion units and RES are presented in Table A7. The results for the storages in Table A8.

Table A7. Equivalent periodic CExC for conversion units and RES technologies.

	Equivalent Periodic CExC $\frac{\text{MW h}}{\text{MW a}}$
Gas boiler	6.83
Gas CHP	24.33
Resistance heater	1.30
Heat pump	2.60
Wind	67.06
PV	347.6
Biomass boiler	8.13
Biomass CHP	81.50
PEM fuel cell	126.68
PEM electrolyser	126.68

Table A8. Equivalent periodic CExC for storage technologies.

	Equivalent Periodic CExC $\frac{\text{MW h}}{\text{MW h a}}$
Battery storage	16.42
Thermal energy storage	0.42
Hydrogen storage	1.24

References

1. Geyer, R.; Knöttner, S.; Diendorfer, C.; Drexler-Schmid, G. *IndustRiES: Energieinfrastruktur für 100% Erneuerbare Energie in der Industrie*; Klima- und Energiefonds der österreichischen Bundesregierung: Wien, Austria, 2019.
2. Sejkora, C.; Kienberger, T. Dekarbonisierung der Industrie mithilfe elektrischer Energie? In Proceedings of the 15. Symposium Energieinnovation, Graz, Austria, 14–16 February 2018; Technische Universität Graz, Ed.; Technische Universität Graz: Graz, Austria, 2018. .
3. Wall, G. Exergy: A Useful Concept. Ph.D. Thesis, Chalmers Tekniska Högskola. N.S, Göteborg, Sweden, 1986; Volume 587.
4. Wall, G. Exergy tools. *Proc. Inst. Mech. Eng. Part A: J. Power Energy* **2003**, *217*, 125–136, doi:10.1243/09576500360611399.
5. Dewulf, J.; van Langenhove, H.; Muys, B.; Bruers, S.; Bakshi, B.R.; Grubb, G.F.; Paulus, D.M.; Sciubba, E. Exergy: Its Potential and Limitations in Environmental Science and Technology. *Environ. Sci. Technol.* **2008**, *42*, 2221–2232, doi:10.1021/es071719a.
6. *Nutzenergieanalyse (NEA) 2017*; Statistik Austria, Ed.; Statistik Austria: Wien, Austria, 2017.
7. Haas, R.; Panzer, C.; Resch, G.; Ragwitz, M.; Reece, G.; Held, A. A historical review of promotion strategies for electricity from renewable energy sources in EU countries. *Renew. Sustain. Energy Rev.* **2011**, *15*, 1003–1034, doi:10.1016/j.rser.2010.11.015.
8. International Energy Agency *Key World Energy Statistics*; Also available on Smartphones and Tablets.; International Energy Agency: Paris, France, 2017.
9. Kåberger, T. Progress of renewable electricity replacing fossil fuels. *Glob. Energy Interconnect.* **2018**, *1*, 48–52, doi:10.14171/j.2096-5117.gei.2018.01.006.
10. Mancarella, P. MES (multi-energy systems): An overview of concepts and evaluation models. *Energy* **2014**, *65*, 1–17, doi:10.1016/j.energy.2013.10.041.
11. Kriechbaum, L.; Scheiber, G.; Kienberger, T. Grid-based multi-energy systems—modelling, assessment, open source modelling frameworks and challenges. *Energy, Sustain. Soc.* **2018**, *8*, 244, doi:10.1186/s13705-018-0176-x.
12. Böckl, B.; Greiml, M.; Leitner, L.; Pichler, P.; Kriechbaum, L.; Kienberger, T. HyFlow—A Hybrid Load Flow-Modelling Framework to Evaluate the Effects of Energy Storage and Sector Coupling on the Electrical Load Flows. *Energies* **2019**, *12*, 956, doi:10.3390/en12050956.

13. Pfenninger, S.; Hawkes, A.; Keirstead, J. Energy systems modeling for twenty-first century energy challenges. *Renew. Sustain. Energy Rev.* **2014**, *33*, 74–86, doi:10.1016/j.rser.2014.02.003.

14. Szargut, J.; Morris, D.R. Cumulative exergy consumption and cumulative degree of perfection of chemical processes. *Int. J. Energy Res.* **1987**, *11*, 245–261, doi:10.1002/er.4440110207.

15. Rant, Z. Exergie - ein neues Wort für die Technische Arbeitsfähigkeit. *Forsch. Auf Dem Geb. Ingenieurwesens 22* **1956**, *22*, 36–37.

16. Krause, T.; Kienzle, F.; Art, S.; Andersson, G. Maximizing exergy efficiency in multi-carrier energy systems. In Proceedings of the Energy Society General Meeting, Providence, RI, USA, 25–29 July 2010; pp. 1–8, doi:10.1109/PES.2010.5589999.

17. Tsatsaronis, G. Definitions and nomenclature in exergy analysis and exergoeconomics. *Energy* **2007**, *32*, 249–253, doi:10.1016/j.energy.2006.07.002.

18. Valero, A.; Lozano, M.A.; Munoz, M. A general theory of exergy saving: I. On the exergetic cost. In *Computer-Aided Engineering of Energy Systems: Vol 3. Second Law Analysis and Modelling, Proceedings of the Winter Annual Meeting of the American Society Mechan. Engin., Anaheim, California, 7–12 December 1986*; Gaggioli, R.A., Ed.; ASME: New York, USA, 1986.

19. Lozano, M.A.; Valero, A. Theory of the exergetic cost. *Energy* **1993**, *18*, 939–960, doi:10.1016/0360-5442(93)90006-Y.

20. Tsatsaronis, G. Thermoeconomic analysis and optimization of energy systems. *Prog. Energy Combust. Sci.* **1993**, *19*, 227–257, doi:10.1016/0360-1285(93)90016-8.

21. Sciubba, E. Beyond thermoeconomics? The concept of Extended Exergy Accounting and its application to the analysis and design of thermal systems. *Exergy, Int. J.* **2001**, *1*, 68–84, doi:10.1016/S1164-0235(01)00012-7.

22. Szargut, J. Towards a rational evaluation of steam prices. *Gospod. Cieptaa* **1957**, *5*, 104–106.

23. Sciubba, E. Exergy-based ecological indicators: From Thermo-Economics to cumulative exergy consumption to Thermo-Ecological Cost and Extended Exergy Accounting. *Energy* **2019**, *168*, 462–476, doi:10.1016/j.energy.2018.11.101.

24. Ziębik, A.; Gładysz, P. Analysis of the cumulative exergy consumption of an integrated oxy-fuel combustion power plant. *Arch. Thermodyn.* **2013**, *34*, 105–122, doi:10.2478/aoter-2013-0018.

25. Wang, S.; Liu, C.; Liu, L.; Xu, X.; Zhang, C. Ecological cumulative exergy consumption analysis of organic Rankine cycle for waste heat power generation. *J. Clean. Prod.* **2019**, *218*, 543–554, doi:10.1016/j.jclepro.2019.02.003.

26. Valero, A.; Lozano, M.A.; Serra, L.; Torres, C. Application of the exergetic cost theory to the CGAM problem. *Energy* **1994**, *19*, 365–381, doi:10.1016/0360-5442(94)90116-3.

27. Valero, A.; Lozano, M.A.; Serra, L.; Tsatsaronis, G.; Pisa, J.; Frangopoulos, C.; von Spakovsky, M.R. CGAM problem: Definition and conventional solution. *Energy* **1994**, *19*, 279–286, doi:10.1016/0360-5442(94)90112-0.

28. Lozano, M.A.; Valero, A. Application of the exergetic costs theory to a steam boiler in a thermal generating station. In *Analysis and Design of Advanced Energy Systems: Applications, Proceedings of the Winter Meeting of the American Society of Mechanical Engineers, Boston, MA, USA, 13–18 December 1987*; American Society of Mechanical Engineers. Winter Annual Meeting. American Society of Mechanical Engineers. Advanced Energy Systems Division. AES (Series); Moran, M.J., Stecco, S.S., Reistad, G.M., Eds.; ASME: New York, USA, 1987.

29. Misra, R.D.; Sahoo, P.K.; Gupta, A. Application of the exergetic cost theory to the LiBr/H_2O vapour absorption system. *Energy* **2002**, *27*, 1009–1025, doi:10.1016/S0360-5442(02)00065-8.

30. Ertesvåg, I.S. Society exergy analysis: A comparison of different societies. *Energy* **2001**, *26*, 253–270, doi:10.1016/S0360-5442(00)00070-0.

31. Ukidwe, N.U.; Bakshi, B.R. Industrial and ecological cumulative exergy consumption of the United States via the 1997 input–output benchmark model. *Energy* **2007**, *32*, 1560–1592, doi:10.1016/j.energy.2006.11.005.

32. Sun, B.; Nie, Z.; Gao, F. Cumulative exergy consumption (CExC) analysis of energy carriers in China. *Int. J. Exergy* **2014**, *15*, 196, doi:10.1504/IJEX.2014.065646.

33. Causone, F.; Sangalli, A.; Pagliano, L.; Carlucci, S. An Exergy Analysis for Milano Smart City. *Energy Procedia* **2017**, *111*, 867–876, doi:10.1016/j.egypro.2017.03.249.

34. Torío, H.; Schmidt, D. Development of system concepts for improving the performance of a waste heat district heating network with exergy analysis. *Energy Build.* **2010**, *42*, 1601–1609, doi:10.1016/j.enbuild.2010.04.002.

35. Li, H.; Svendsen, S. Energy and exergy analysis of low temperature district heating network. *Energy* **2012**, *45*, 237–246, doi:10.1016/j.energy.2012.03.056.

36. Bösch, M.E.; Hellweg, S.; Huijbregts, M.A.J.; Frischknecht, R. Applying cumulative exergy demand (CExD) indicators to the ecoinvent database. *Int. J. Life Cycle Assess.* **2007**, *12*, 181–190, doi:10.1065/lca2006.11.282.

37. Hilpert, S.; Kaldemeyer, C.; Krien, U.; Günther, S.; Wingenbach, C.; Plessmann, G. The Open Energy Modelling Framework (oemof)—A new approach to facilitate open science in energy system modelling. *Energy Strategy Rev.* **2018**, *22*, 16–25, doi:10.1016/j.esr.2018.07.001.

38. Oemof Developer Group. *Oemof—Open Energy Modelling Framework (v0.2.2)*; Oemof Developer Group: Berlin, Germany, 2018 ; doi:10.5281/zenodo.1302372.

39. United Nations. (Ed.) *International Recommendations for Energy Statistics (IRES)*; Statistical Paper. Series M; United Nations: New York, NY, USA, 2017.

40. Böckl, B.; Kriechbaum, L.; Kienberger, T. Analysemethode für kommunale Energiesysteme unter Anwendung des zellularen Ansatzes. In *14. Symposium Energieinnovation*; Institut für Elektrizitätswirtschaft und Energieinnovation, Ed.; TU Graz: Graz, Austria, 2016.

41. Kleinertz, B.; Pellinger, C.; von Roon, S.; Hübner, T.; Kaestle, G. *EU Displacement Mix: A Simplified Marginal Method to Determine Environmental Factors for Technologies Coupling Heat and Power in the European Union*; Forschungstelle für Energiewirtschaft eV: Munich, Germany, 2018.

42. Frischknecht, R.; Stucki, M.; Flury, K.; Itten, R.; Tuchschmid, M. *Primärenergiefaktoren von Energiesystemen: Version 2.2*; ESU-Services Ldt.: Uster, Switzerland, 2012.

43. oemof Developer Group. *The Oemof Demandlib (Oemof. Demandlib)*; oemof Developer Group: Berlin, Germany, 2016; doi:10.5281/ZENODO.438786.

44. Pfenninger, S.; Staffell, I. Long-term patterns of European PV output using 30 years of validated hourly reanalysis and satellite data. *Energy* **2016**, *114*, 1251–1265, doi:10.1016/j.energy.2016.08.060.

45. Staffell, I.; Pfenninger, S. Using bias-corrected reanalysis to simulate current and future wind power output. *Energy* **2016**, *114*, 1224–1239, doi:10.1016/j.energy.2016.08.068.

46. Hellwig, M. Entwicklung und Anwendung parametrisierter Standard-Lastprofile. *Ph.D. Thesis*, Technische Universität München, München, Germany, 2003.

47. BDEW/VKU/GEODE. *BDEW/VKU/GEODE—Leitfaden: Abwicklung von Standardlastprofilen Gas*; BDEW/VKU/GEODE: Berlin, Germany, 2015.

48. E-Control. *Sonstige Marktregeln Strom*; Kapitel 6 - Zählwerte, Datenformate und standardisierte Lastprofile; E-Control: Wien, Austria, 2012.

49. Probas. Holz-EU-KUP-Pellet-Heizung-50 kW-2030. Available online: https://www.probas.umweltbundesamt.de/php/prozessdetails.php?id=%7B4D4E8B6D-1FEC-45E5-B927-7977C6455186%7D (accessed on 17 October 2019).

50. Probas. Gas-Heizung-Brennwert-DE-2030. Available online: https://www.probas.umweltbundesamt.de/php/prozessdetails.php?id=%7BE949C50B-659D-42C4-A779-2FCCD9CCA9AF%7D (accessed on 17 October 2019).

51. Probas. el.heat pump-mono-CZ. Available online: https://www.probas.umweltbundesamt.de/php/prozessdetails.php?id=%7BCC0E4B28-80DA-11D4-9E81-0080C8426C9A%7D (accessed on 17 October 2019).

52. Ausfelder, F.; Dura, H.E. (Eds.) *Optionen für ein nachhaltiges Energiesystem mit Power-to-X Technologien: Herausforderungen - Potenziale - Methoden - Auswirkungen : 1. Roadmap des Kopernikus-Projektes "Power-to-X": Flexible Nutzung erneuerbarer Ressourcen (P2X) : Erstellt im Rahmen der Roadmapping-Aktivitäten im Koperinkus-Projekt "Power-to-X": Flexible Nutzung erneuerbarer Ressourcen (P2X) gefördert durch das Bundesministerium für Bildung und Technologie*, 1st ed.; DECHEMA Gesellschaft für Chemische Technik und Biotechnologie e.V: Frankfurt am Main, Germany, 31 August 2018.

53. Probas. el.heating-CZ. Available online: https://www.probas.umweltbundesamt.de/php/prozessdetails.php?id=%7BCC0E4B29-80DA-11D4-9E81-0080C8426C9A%7D (accessed on 17 October 2019).

54. Probas. Holz-HS-Waldholz-HKW-DM-DE-2005/brutto. Available online: https://www.probas.umweltbundesamt.de/php/prozessdetails.php?id=%7B67221E5B-47FF-45D8-B8E2-08DF837417C3%7D (accessed on 17 October 2019).

55. Probas. Gas-BHKW-250-DE-2010/en. Available online: https://www.probas.umweltbundesamt.de/php/prozessdetails.php?id=%7B4C6CB40B-4922-4648-9922-31CE39A826C0%7D (accessed on 17 October 2019).

56. Peters, J.F.; Baumann, M.; Zimmermann, B.; Braun, J.; Weil, M. The environmental impact of Li-Ion batteries and the role of key parameters—A review. *Renew. Sustain. Energy Rev.* **2017**, *67*, 491–506, doi:10.1016/j.rser.2016.08.039.

57. heizungsdiscount24.de. 10000 Liter TWL Puffer, Pufferspeicher Typ P10000. Available online: https://www.heizungsdiscount24.de/speichertechnik/10000-liter-twl-puffer-pufferspeicher-typ-p10000.html (accessed on 17 October 2019).

58. Deutz, S.; Bongartz, D.; Heuser, B.; Kätelhön, A.; Schulze Langenhorst, L.; Omari, A.; Walters, M.; Klankermayer, J.; Leitner, W.; Mitsos, A.; et al. Cleaner production of cleaner fuels: Wind-to-wheel—environmental assessment of CO_2—based oxymethylene ether as a drop-in fuel. *Energy Environ. Sci.* **2018**, *11*, 331–343, doi:10.1039/C7EE01657C.

59. Probas. Solar-PV-multi-Rahmen-mit-Rack-DE-2030. Available online: https://www.probas.umweltbundesamt.de/php/prozessdetails.php?id=%7B05EC9CDC-9418-45CD-A333-A1CC3CCF7C48%7D (accessed on 17 October 2019).

60. Probas. wind Mill-CZ. Available online: https://www.probas.umweltbundesamt.de/php/prozessdetails.php?id=%7BCC0E4E86-80DA-11D4-9E81-0080C8426C9A%7D (accessed on 17 October 2019).

61. Fratzscher, W.; Brodjanskij, V.; Michalek, K. *Exergie: Theorie und Anwendung*; Springer: Berlin/Heidelberg, Germany, 2013.

62. Probas. MetallStahl-mix-DE-2010. Available online: https://www.probas.umweltbundesamt.de/php/prozessdetails.php?id={F428D8DE-27A7-4FD6-850C-4C0DC39740DE} (accessed on 17 October 2019).

63. Probas. Chem-OrgPVC-mix-DE-2030. Available online: https://www.probas.umweltbundesamt.de/php/prozessdetails.php?id={C0A5A645-F236-4D6F-8B75-24DB0D899824} (accessed on 17 October 2019).

64. Probas. Steine-ErdenZement-DE-2010. Available online: https://www.probas.umweltbundesamt.de/php/prozessdetails.php?id={8BBACD2C-DA5F-43A7-B24B-0D49BA161D26} (accessed on 17 October 2019).

65. Probas. Chem-OrgHDPE-DE-2000. Available online: https://www.probas.umweltbundesamt.de/php/prozessdetails.php?id={0E0B299A-9043-11D3-B2C8-0080C8941B49} (accessed on 17 October 2019).

66. Probas. MetallAluminium-mix-DE-2030. Available online: https://www.probas.umweltbundesamt.de/php/prozessdetails.php?id={5CAE9D2B-4C86-4244-88B2-A79E09DC870B} (accessed on 17 October 2019).

67. Probas. MetallKupfer-DE-mix-2030. Available online: https://www.probas.umweltbundesamt.de/php/prozessdetails.php?id={AB07F312-569F-452A-82B5-3BF124101151} (accessed on 17 October 2019).

68. Probas. FabrikSilizium-Modul-multi-DE-2030. Available online: https://www.probas.umweltbundesamt.de/php/prozessdetails.php?id={C8A517A8-CC5B-4559-9747-E385A6E4FC39} (accessed on 17 October 2019).

 *energies*

Article

Optimising the Concentrating Solar Power Potential in South Africa through an Improved GIS Analysis

Dries. Frank Duvenhage [1,*], Alan C. Brent [1,2], William H.L. Stafford [1,3] and Dean Van Den Heever [4]

1 Engineering Management and Sustainable Systems, Department of Industrial Engineering, the Solar Thermal Energy Research Group and the Centre for Renewable and Sustainable Energy Studies, Stellenbosch University, Stellenbosch 7602, South Africa; alan.brent@vuw.ac.nz (A.C.B.); wstafford@csir.co.za (W.H.L.S.)
2 Sustainable Energy Systems, School of Engineering and Computer Science, Victoria University of Wellington, Wellington 6140, New Zealand
3 Green Economy Solutions, Natural Resources and the Environment, Council for Scientific and Industrial Research, Stellenbosch 7600, South Africa
4 Legal Drone Solutions, Stellenbosch 7600, South Africa; deanvdh@gmail.com
* Correspondence: dfrankduv@gmail.com

Received: 11 May 2020; Accepted: 16 June 2020; Published: 23 June 2020

Abstract: Renewable Energy Technologies are rapidly gaining uptake in South Africa, already having more than 3900 MW operational wind, solar PV, Concentrating Solar Power (CSP) and biogas capacity. CSP has the potential to become a leading Renewable Energy Technology, as it is the only one inherently equipped with the facility for large-scale thermal energy storage for increased dispatchability. There are many studies that aim to determine the potential for CSP development in certain regions or countries. South Africa has a high solar irradiation resource by global standards, but few studies have been carried out to determine the potential for CSP. One such study was conducted in 2009, prior to any CSP plants having been built in South Africa. As part of a broader study to determine the impact of CSP on South Africa's water resources, a geospatial approach was used to optimise this potential based on technological changes and improved spatial data. A tiered approach, using a comprehensive set of criteria to exclude unsuitable areas, was used to allow for the identification of suitable areas, as well as the modelling of electricity generation potential. It was found that there is more than 104 billion m^2 of suitable area, with a total theoretical potential of more than 11,000 TWh electricity generating capacity.

Keywords: concentrating solar power; geographic information systems; potential assessment; renewable energy; solar thermal; South Africa; CSP

---

1. Introduction

Global interests in CSP is said to grow with 87% by 2021 [1], with South Africa likely to undergo various possible development scenarios [2]. The approach applied in literature for determining potential CSP capacities for a region or country typically follows a generic tiered approach using geographical information systems (GIS) [3]. First, certain spatial zones within a region are removed from consideration based on explicit exclusion criteria, while others are regarded as more suitable based on preferred inclusion criteria. Thereafter, considerations are made for distance to and from required supporting infrastructure, such as the electrical transmission network, roads and water sources. Once these limits have been implemented, suitable zones are then identified within the respective region or country.

Second, the CSP potential of these suitable zones are then calculated based on assumptions regarding Land Use Efficiencies (LUE, %) or Power Densities (km^2/GW). Due to the complex nature of multiple criteria being used, these zones can also be ranked based on certain economic, social or technical criteria, according to a multi-criteria decision-making method, such as an analytical hierarchy process (AHP). An AHP makes use of hierarchal structures to represent a problem and incorporates priority scales based on key criteria to inform decision making [4]. Third, and finally, the potentials for generation are then aggregated and/or ranked according to administrative borders, or some other spatially explicit boundaries of interest. Once this has been done, estimations can be made of the amount of potential energy that can be generated, based on further assumptions regarding plant capacity factor and location-specific conditions. This method is illustrated in Figure 1.

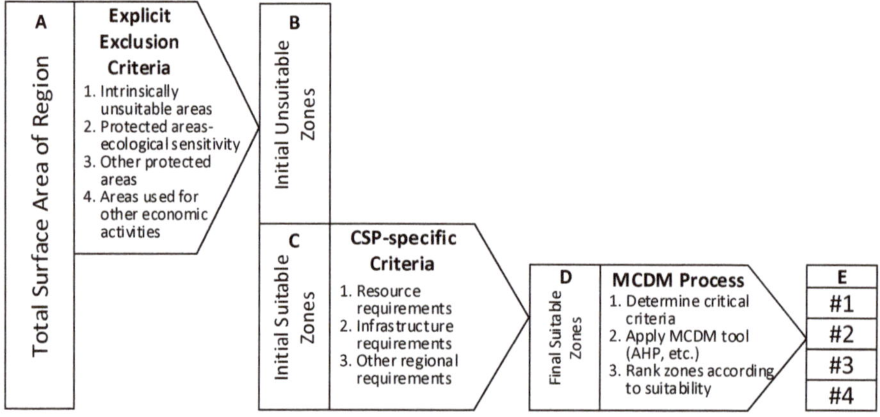

Figure 1. Standard approach for Concentrating Solar Power (CSP) potential studies.

A detailed account of the suitability criteria from various CSP potential studies using the process depicted in Figure 1 is given in [5]. The ranking of suitable zones in step E is optional. Typically, after identifying the final suitable zones in Step D, calculations are done to determine the potential CSP capacity associated with those zones. This, of course, requires that certain assumptions are made regarding the CSP technology; the plant configuration and site-specific solar resource, particularly that of Direct Normal Irradiance (DNI); and meteorological conditions. Finally, as mentioned before, these capacities can be represented as installed power or total annual electricity generated, and are then aggregated according to user-defined discrete borders, such as administrative boundaries of provinces or districts, economic development areas or some other definition of interest.

The aim of this paper is to apply this approach to optimise the theoretical potential of CSP in South Africa, by considering recent technology changes and improved spatial data that is now available.

2. Materials and Methods

Before determining whether a certain area is technically suitable for the development of a CSP plant, all explicitly non-suitable areas are to be excluded. These areas generally constitute "no-go" areas for infrastructure development due to intrinsic unsuitability, or because of a conflict of some sort. In the case of South Africa (SA), these areas can be grouped into intrinsic unsuitability, ecological and economic conflict.

2.1. Exclusion Criteria: Intrinsically Unsuitable Areas

As CSP requires large areas of reflective mirror surfaces, typically unsuitable areas include surface water (rivers and dams) and steep slopes. Rivers can be classified based on how their water flows vary seasonally (class) and what order they are within a catchment (1 to 7). The class can be perennial, non-perennial or dry, while the order refers to how many river-branches have joined upstream of a certain river-segment. For example, if Streams A and B converge to form Stream C, but Streams A and B have no converging streams upstream of them, then Streams A, B and C have orders of 1, 1 and 2, respectively. These two classifications are important when deciding which streams can easily be diverted in large, flat catchments and which cannot. Therefore, streams which adhere to the following classifications have been excluded: (Class = Perennial), AND (Order > = 5). This implies that dry river beds in large flat areas, for example, are considered technically suitable areas, while large perennially flowing rivers with many upstream converging rivers are not. An arbitrary buffer of 5 m was applied to the rivers data file. All dams are excluded, with a 500 m buffer around their polygons. The data files for these exclusion criteria are from the Department of Water and Sanitation of SA, and are available online. Figure 2A shows the rivers and dams.

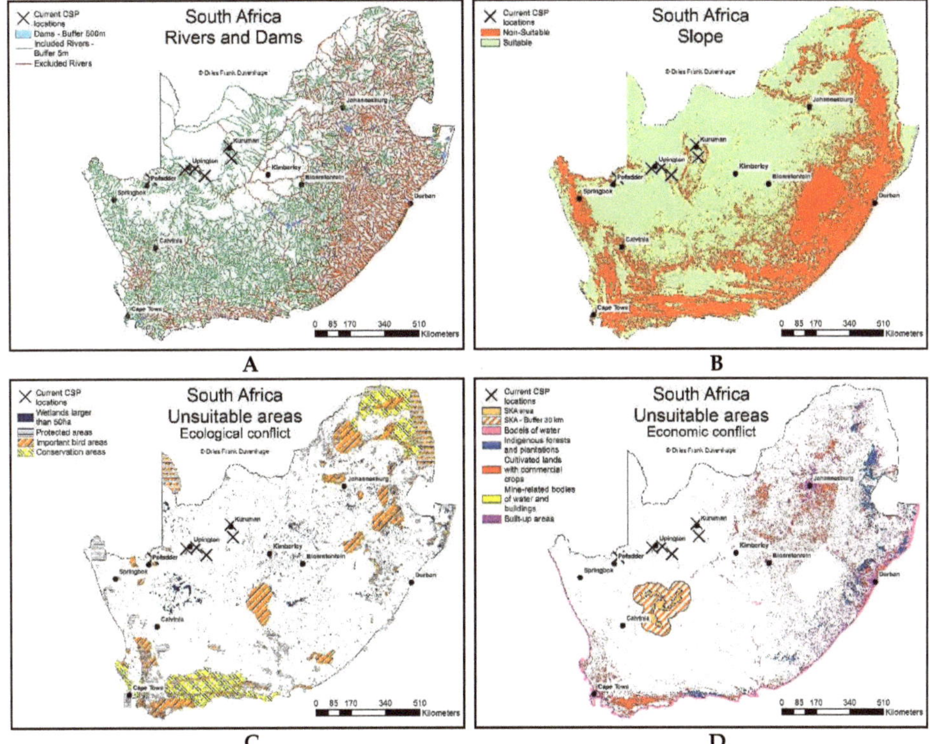

Figure 2. Maps of South Africa showing excluded areas due to (**A**) rivers and dams, (**B**) slope, (**C**) unsuitable areas—ecological conflict and (**D**) unsuitable areas—economic conflict.

The second most critical technical feasibility criterion for a CSP plant is whether it is physically possible to construct the plant on the ground. The major criterion is slope, or the rate of change in vertical altitude over a certain horizontal distance. This can be expressed in degrees or % change in altitude per horizontal distance. The two existing commercial CSP technologies—Parabolic Troughs (PT) and Central Receivers (CR)—are capable of being constructed on surfaces that have a slope of between 1% (0.57°) and 7% (4.00°), relating to a rise of between 1 m and 7 m over a horizontal distance of 100 m [6]. Surfaces with slopes steeper than this range would require intensive civil works and ground preparation to construct the long rows of mirrors required for PT or complicate the construction and layout of the thousands of heliostats used to reflect sunlight onto the receiver mounted on a high tower, in the case of CR. The slope used in this study is 3%, in accordance with the slope found at existing CSP sites in SA, as shown in Table 1, and with the previous study for South Africa [7]. A digital elevation model raster file, showing height above sea level, from the South African Environmental Observation Network, was used, with a spatial resolution of 90 m × 90 m. A slope raster, using the ESRI ArcGIS®® (ESRI, Redlands, CA, USA) Spatial analyst slope toolbox, was generated, shown in Figure 2B.

Table 1. Suitability conditions at existing CSP plants in SA.

CSP Plant Name	CSP + Cooling Type	Location (Decimal Degrees)	DNI (kWh/m^2/y)	Slope (%)	Km to Excl. Water Body	Km to Ecological Conflict Area *	Km to Economic Conflict Area	Km to Tx Liné **
Kaxu and Xina	PTDC	−28.89, 19.59	2927	1.17	<27	<1 (IBA)	<26	<4 (Ex)
Khi Solar One	CRDC	−28.54, 28.08	2922	0.58	<11	<12 (wet)	<8	<4 (Pl)
Kathu Solar Park	PTDC	−27.61, 23.04	2801	0.44	<50	<0.5 (wet)	<7	<9 (Ex)
Bokpoort	PTWC	−28.78, 21.96	2930	2.44	<2	<5 (wet)	<5	<0.4 (Pl)
Illanga	PTDC	−28.49, 21.52	2912	2.44	<12	<0.3 (wet)	<11	<31 (Pl)
This study	NA	NA	>2400	<3.00	5 m	NA	NA	<20

* IBA: important bird areas, wet: Wetalnds; ** Ex: Existing Tx lines, Pl: Planned Tx lines.

2.2. Exclusion Criteria: Unsuitable Areas due to Ecological Conflict

Regarding CSP development, the same basic ecological exclusion criteria apply as to any large infrastructure project. In SA, the following national areas are always considered "no-go" areas: important bird and biodiversity areas, conservation areas, protected areas and wetlands.

Conservation and protected areas are determined by the national Department of Environmental Affairs under the Protected Areas Act of 2003. These databases contain areas under formal legislative protection, of which the legal statuses of these areas are audited against official gazettes before inclusion. The inclusion of certain areas into the database is governed by the relevant environmental conventions of the Act and is updated at quarterly intervals. The types of areas considered as conservation or protected include biosphere reserves, botanical gardens, wetlands, forest nature reserves, forest wilderness areas, marine protected areas, mountain catchments, national parks, nature reserves, protected and special nature reserves and world heritage sites. For wetlands, only those larger than 0.5 km^2 were excluded. Important bird areas (IBA), compiled by Birdlife South Africa, are objectively determined using globally accepted criteria. An IBA is selected based on the presence of the following bird categories: bird species of global or regional conservation concern, assemblages of restricted range bird species, assemblages of biome-restricted bird species and concentrations of congregatory bird species [8]. These unsuitable areas are shown in Figure 2C.

2.3. Exclusion Criteria: Unsuitable Areas due to Economic Conflict

Another basis on which any new large-scale infrastructure development can be excluded from consideration in a particular area is the likely conflict with existing economic activities. In SA, agriculture uses a significant portion of land to support many people's livelihoods and contribute to the economy. In order to determine which areas are not suited to CSP plants because of a conflicting economic activity, the 72-class 2013–2014 South African National Land-Cover Dataset was used, which is a 30 × 30 m raster for the entire SA. This dataset was compiled by Geoterraimage for the Department of Environmental Affairs [9] Since the list of excluded classes totals 59, only a brief overview of these shall be given, and readers are invited to review the relevant document, available online, for more detail. Naturally, areas classified as bodies of water were excluded (class 1–2). Indigenous forests were also excluded since special environmental permits are required to clear these for infrastructure developments (class 4). Cultivated lands with commercial annual rainfed and irrigated crops, commercial permanent crops, and commercial sugarcane crops were excluded due to their high economic value (classes 10, 11, 13–17, 19, 20, 22, 26–31). All forest plantations were excluded (classes 32–34), as well as mine-related bodies of water and buildings (classes 37–39). Finally, all built-up areas were excluded due to conflict with human settlements (classes 42–72). Another type of area excluded from consideration is the area dedicated to the construction and development of the Square Kilometer Array (SKA). This is an international project to build the world's largest radio telescope, with eventually over a million square meters of collecting area. The area within SA which has formally been dedicated to the SKA, is located in the Northern Cape Province, and is shown in Figure 2D, along with the other excluded areas from the land cover dataset. This area includes a buffer of 30 km for electrical infrastructure with a rating of greater than 100 kVA, to prevent interference with the sensitive radio telescope equipment [10].

2.4. CSP-Specific Suitability Criteria: DNI

The most critical suitability criterion for the determination of CSP potential is Direct Normal Irradiance (DNI), measured in kWh/m^2 per time period (day, month or year). This is the amount of direct irradiance incident on a surface normal to the direction of the sunrays. The amount of DNI which reaches the earth's surface is influenced by the relative position of the earth to the sun (season and time of day), position on the surface of the earth (elevation, latitude and longitude) and atmospheric conditions (aerosols, dust, water vapour and most importantly clouds) [11]. As DNI is the primary energy source for a CSP plant, it is the major determining factor of a plant's techno-economic feasibility. For this reason, the general consensus is that the minimum required DNI is between 1800 and 2000 kWh/m^2/y, with increases in DNI directly related to reductions in cost of generated electricity [12]. The threshold DNI employed in any CSP potential study is relative to the average DNI in the study area. The inclusion of areas with lower DNI, even though it might be equal to or larger than 1800 kWh/m^2/y, will only result in the consideration of areas that are less favourable than others. SA has a minimum, average and maximum DNI of 1290 kWh/m^2/y, 2397 kWh/m^2/y and 3141 kWh/m^2/y, respectively. The annual DNI data source used in this study is from SolarGIS ®®, and is a raster file with a spatial resolution of 30 arc-seconds, shown in Figure 3A [13]. Based on the high average DNI of SA, the DNI threshold selected in this study is 2400 kWh/m^2/y, as opposed to the minimum of 2000 kWh/m^2/y used in other study areas with substantially lower average DNI.

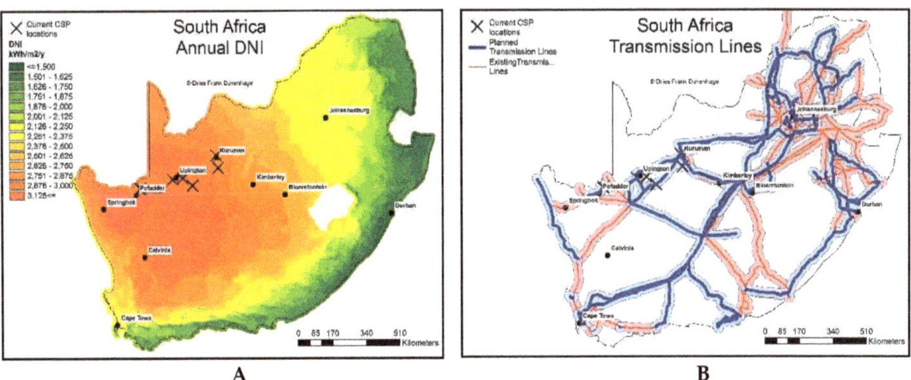

A   **B**

Figure 3. Maps of South Africa showing CSP-related suitability criteria (**A**) Direct Normal Irradiance (DNI) and (**B**) transmission network.

2.5. CSP-Specific Suitability Criteria: Distance to Required Infrastructure

Theoretically, CSP infrastructure only requires suitable land and high DNI to be feasible. However, for this infrastructure to be economically and logistically practical, it must be located near adequate transport, transmission and water supply infrastructure. In fact, water is the only other natural resource on which CSP depends [3]. The co-location of theoretically suitable areas and artificial infrastructure can be directly due to the development of the CSP infrastructure, or due to independent, existing infrastructure expansion plans. In the case of the former, the costs associated with these infrastructure developments must be added to those of the CSP plant, but not necessarily in the case of the latter.

The same distance to transmission (Tx) infrastructure will be used as in [7], namely less than, or equal to, 20 km. The use of 20 km as the limit for distance to transmission infrastructure is to ensure continuity with previous the study done for CSP potential in South Africa. Ultimately, the selection of this limit should be based on careful techno-economic modelling of the selected CSP technology configuration's performance at a particular location, and its resulting profit (a function of performance and power purchase tariffs) and weighed against the cost of transmission network connection. These considerations are, however, beyond the scope of this study. This study does, however, expand on these criteria by comparing it to that of actual CSP plants built in South Africa (Table 1), and by the inclusion of Planned transmission lines, as indicated by the national, fully vertically integrated electricity supplier, Eskom.

This study did not include consideration for transport infrastructure, as dirt roads are a low-cost option for accessing highly suitable areas. Furthermore, no consideration has been given to proximity to water infrastructure, as was the case in [7], and is not included here, although it is the focus of further research efforts [14]. This work includes new planned Tx lines in its analysis to explore likely future CSP potential in these areas. The transmission network in SA can be seen in Figure 3B, acquired from the national utility, Eskom.

2.6. Suitability Conditions at Existing CSP Plants in SA

To align the assumptions used in this work with the actual conditions at existing CSP plants in SA, the available GIS datasets were used to evaluate the locations of these plants. The results are summarized in Table 1 for PT with wet cooling (PTWC) and dry-cooling (PTDC), and CR with wet-cooling (CRWC) and dry-cooling (CRDC).

The DNI at all locations of existing operational CSP plants in South Africa is above 2800 kWh/m^2/y, compared to the generally accepted minimum of 2000 kWh/m^2/y. The slope at these sites varies between 0.4% and 2.5%, therefore the use of a maximum acceptable slope of 3% in this study. Generally, the sites are located relatively far from economic exclusion areas. It is apparent that the sites are also mostly near to high-voltage Tx lines, and that some of the plants have been built near to those considered "under planning", indicating that these lines might already have existing by the time the plants were connected to the grid.

3. Results

The final aim of the greater project, which this work forms part of, is to model monthly CSP operation and water consumption across large geographical areas. To simplify this, SA was divided into a 1 × 1 km grid. Blocks within the grid were then excluded if more than 50% of their area (0.5 km^2) intersected with any of the explicit shapefiles of the various exclusion criteria. The exclusion reason(s) for a certain block was recorded as attribute data in that block. The result was a grid leaving only suitable blocks, either located near Existing or Planned Tx lines, shown in Figure 4A,B, respectively. Thereafter, the DNI values intersecting with these suitable areas were stored as attribute data in each block.

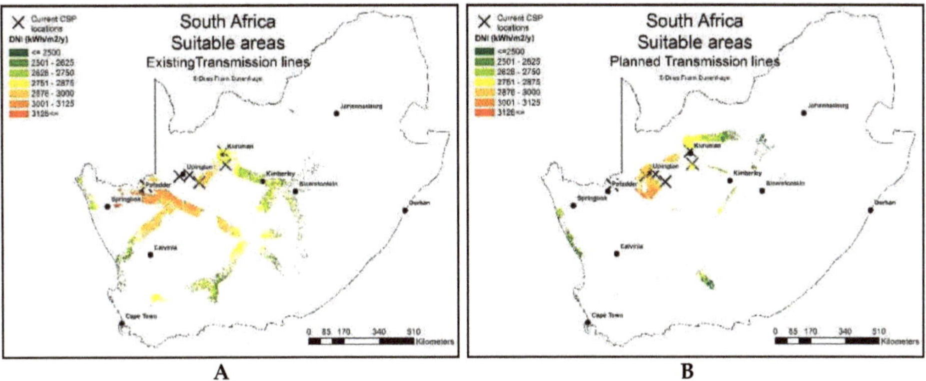

A   **B**

Figure 4. Maps of South Africa showing suitable areas near (**A**) existing Tx lines and (**B**) planned Tx lines.

CSP Potential Modelling

The potential amount of electrical energy generated from a CSP plant in a specific location depends on the CSP technology configuration in use and the solar resources at that location. The CSP technology configuration used depends greatly on the business model used to generate profit, the tariff structure and various financially driven criteria. However, to determine the theoretical potential at a location, certain high-level assumptions can be made based on the fundamental energy conversions taking place in a CSP plant, and these assumptions can be used to estimate the amount of electricity, in GWh, a CSP plant can generate, based on knowledge of the prospective location's DNI. A common approach is using efficiencies for the energy conversions in a CSP plant. This process was explained in detail, with relevant equations and assumptions, in [5]. For clarity, the governing equations are given in Equation (1).

$$Q_{ELNET} = LUF \times N_{SE} \times DNI \times A, \tag{1}$$

Q_{ELNET} represents the net electrical energy (GWh) that is generated, based on the Land Use Factor (*LUF*) and area (*A*). *LUF* refers to the ratio between total footprint and solar field area. The assumptions used in this work, to calculate total annual generation potential, are given in Table 2. The Solar-to-electric efficiency (STE) values were determined by modelling the annual operation of a 50 MW PTWC, PTDC, CRWC and CRDC plant, with 9 h of storage at the locations of the five existing CSP plants in SA [15]. A storage capacity of 9 h was arbitrarily selected based on the maximum storage capacity of an operational CSP plant in South Africa (Bokpoort). This value has minimal effect on the annual STE of a CSP plant. Furthermore, if a simplified STE is used to calculate electrical energy generation potential, the thermal storage required can be calculated if a specific installed generation capacity is specified. The annual average STE depends greatly on seasonal variations in DNI, solar geometry and atmospheric conditions (temperature and relative humidity). Monthly simulations were therefore carried out at each of the locations identified and summed over a period of 12 months to determine annual results, as explained in detail in [15].

Table 2. Efficiency assumptions used.

Parameter Description	Symbol	Unit	PTWC	PTDC	CRWC	CRDC
Land Use Factor	*LUF*	%	28	28	23	23

4. Conclusions

This paper applied a standard approach to optimise the theoretical potential of CSP in South Africa and showed how recent technology changes and improved spatial data need to be considered for a comprehensive analysis. An expanded version of Equation 1 was used to calculate the monthly generation potential from the four different CSP and cooling technology configurations for each block in the suitable areas grid. This study calculated the potential for both PT and CR CSP technologies, as well as in combination with wet- or dry-cooling, to reflect the impact that such design decisions will have. In Figure 5, the annual generation potential in GWh per 1 × 1 km block is shown for the wet-cooled options for PT and CR, near existing and planned Tx lines, respectively. The total suitable area identified is 104,709 km^2, which is less than 9% of the total SA surface area.

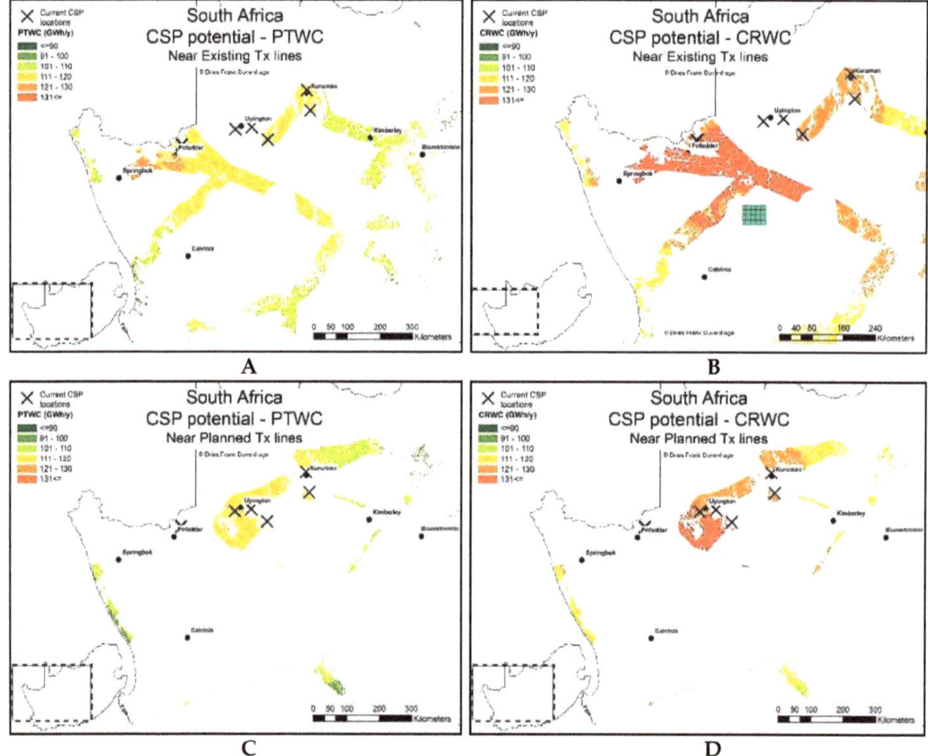

Figure 5. Maps of South Africa showing Annual generation potential from CSP: (**A**) PT with wet cooling (PTWC) near existing Tx lines, (**B**) CR with wet-cooling (CRWC) near existing Tx lines, (**C**) PTWC near planned Tx lines and (**D**) CRWC near planned Tx lines.

The higher efficiency of CR plants results in the generally higher annual electricity generation reflected in the colour scale. The quantitative results are given in Table 3. The annual generation potential from CSP at all suitable locations identified in this study exceeds the national annual demand for electricity in South Africa by between 4.768% and 5.184%. In fact, it would require less than 2400 km^2, or 0.2% of the total SA surface area (2% of the identified suitable areas) with the highest DNI, covered with the least efficient PTDC configuration to generate the annual electricity demand of 231.5 TWh for 2018 [16]. This surface area is shown in Figure 5B, for spatial reference, and depends on the DNI of the locations selected, but serves as a general indication of relative size.

Table 3. Summary of CSP potential results in identified suitable areas.

Transmission Line	Area (km^2)	PTWC (TWh/y)	PTDC (TWh/y)	CRWC (TWh/y)	CRDC (TWh/y)
Existing	71,457	7695		8191	8069
Planned	33,252	3581	3505	3811	3754
Total	104,709	11,276	11,037	12,002	11,823
% SA Total *	8.59%	4871%%	4768%	5107%	5184%

* SA area: 1,221,037 km^2. SA 2018 electricity demand: 231.5 TWh, [16].

To put this into perspective, the total amount of electricity generated from all RETs in South Africa in 2018 was 10.483 TWh, less than 4.53% of the 2018 annual electricity demand [17]. The major advantage of CSP, in comparison to PV and wind without storage, is that it is inherently capable of large-scale storage for dispatchable generation. This is a critical benefit over these RETs, as the energy generated from CSP can coincide with the times of day when it is needed from the grid, not only when the resource is available, as is the case for PV and wind. However, in light of the recent decline in battery storage costs (−85% compared to 2010 values), and cost of generation from PV and wind (−85% and −49% compared to 2010 values, respectively), a more detailed economic evaluation of CSP generation potential in SA is needed to fairly compare it with other RET and storage options [18].

However, the value that CSP can hold for longer duration energy storage and dispatch in countries with high DNI generation potential, like that pointed out in this study for South Africa, should be included in planning models. This is of particular importance when planning the replacement of fossil-fuel based dispatchable energy (coal, diesel and natural gas) in the electricity supply mix, and to assist in transitioning to a cleaner energy mix dominated by renewables.

Author Contributions: All authors have read and agree to the published version of the manuscript. Conceptualization, D.F.D.; methodology, D.F.D.; software, D.V.D.H.; validation, D.F.D.; formal analysis, D.F.D.; investigation, D.F.D.; data curation, D.V.D.H.; writing—original draft preparation, D.F.D.; writing—review and editing, A.C.B. and W.H.L.S.; visualization, D.F.D.; supervision, A.C.B. and W.H.L.S. project administration, A.C.B.

Funding: This research received no external funding.

Acknowledgments: This work was partially funded by Stellenbosch University's Solar Thermal Energy Research Group and Industrial Engineering Department. Thanks go out to the various data sources, as referenced. All glory to God.

Conflicts of Interest: The authors declare no conflicts of interest.

References

1. IEA. *World Energy Outlook 2018*; IEA: Paris, France, 2018. Available online: https://www.iea.org/reports/world-energy-outlook-2018 (accessed on 18 March 2019).
2. Duvenhage, D.F.; Craig, O.O.; Brent, A.C.; Stafford, W.H.L. Future CSP in South Africa—A review of generation mix models, their assumptions, methods, results and implications. *AIP Conf. Proc.* **2018**, *2033*, 120002. [CrossRef]
3. Duvenhage, D.F.; Brent, A.C.; Stafford, W.H.L. The need to strategically manage CSP fleet development and water resources: A structured review and way forward. *Renew. Energy* **2019**, *132*, 813–825. [CrossRef]
4. Ziuku, S.; Seyitini, L.; Mapurisa, B.; Chikodzi, D.; Van Kuijk, K. Potential of Concentrated Solar Power (CSP) in Zimbabwe. *Energy Sustain. Dev.* **2014**, *23*, 220–227. [CrossRef]
5. Duvenhage, D.F.; Brent, A.C.; Stafford, W.H.L.; Craig, O.O. Water And CSP—A Preliminary Methodology for Strategic Water Demand Assessment. *AIP Conf. Proc.* **2019**, *2126*, 220002. [CrossRef]
6. Ramdé, E.W.; Azoumah, Y.; Brew-Hammond, A.; Rungundu, A.; Tapsoba, G. Site Ranking and Potential Assessment for Concentrating Solar Power in West Africa. *Nat. Resour.* **2013**, *04*, 146–153. [CrossRef]
7. Fluri, T.P. The potential of concentrating solar power in South Africa. *Energy Policy* **2009**, *37*, 5075–5080. [CrossRef]
8. Marnewick, M.; Retief, E.; Theron, N.; Wright, D.; Anderson, T. BirdLife South Africa. Important Bird and Biodiversity Areas of South Africa: Johannesburg, South Africa, 2015.

9. GEOTERRAIMAGE. *2013–2014 South African National Land Land-Cover Dataset*; GEOTERRAIMAGE: Pretoria, South Africa, 2015; Volume 2.

10. Government Gazette. *Regulations On The Protection Of The Karoo Central Astronomy Advantage Areas in Terms Of The Astronomy Geographic Advantage Act, 2007*; Government Gazette: Pretoria, South Africa, 2017; pp. 171–226.

11. Sengupta, M.; Habte, A.; Kurtz, S.; Dobos, A.; Wilbert, S.; Lorenz, E.; Stoffel, T.; Renné, D.; Gueymard, C.; Myers, D.; et al. *Best Practices Handbook for the Collection and Use of Solar Resource Data for Solar Energy Applications*; NREL/TP-5D00-63112; NREL: Golden, CO, USA, 2015.

12. Kearney, A.T. ESTELA Solar Thermal Electricity 2025. Available online: https://www.estelasolar.org/portfolio_page/solar-thermal-electricity-2025/ (accessed on 12 November 2018).

13. SolarGIS Solar Resource Maps of South Africa. Available online: https://solargis.com/maps-and-gis-data/download/south-africa (accessed on 22 March 2018).

14. Duvenhage, D.F.; Brent, A.C.; Stafford, W.H.L.; Grobbelaar, S. Water and CSP—Linking CSP water demand models and national hydrology data to sustainably manage CSP development and water resources in arid regions. *Sustainability* **2020**, *12*, 3373. [CrossRef]

15. Duvenhage, D.F. Sustainable Future CSP Fleet Deployment in South Africa: A Hydrological Approach to Strategic Management. Ph.D.Thesis, Stellenbosch University, Stellenbosch, South Africa, 2019. Ph.D.Thesis, Stellenbosch University, Stellenbosch, South Africa, 2019.

16. *Statistics South Africa, Electricity Generated and Available for Distribution (Preliminary)*; Statistics South Africa: Pretoria, South Africa, 2019.

17. IPP Office. *Independent Power Producers Procurement Programme An Overview December 2018*; IPP Office: Pretoria, South Africa, 2018.

18. Bloomberg New Energy Finance. *New Energy Outlook*. 2019. Available online: https://about.bnef.com/new-energy-outlook/#toc-download (accessed on 21 January 2020).

Article

Optimal Operational Strategy for Power Producers in Korea Considering Renewable Portfolio Standards and Emissions Trading Schemes

Dongmin Son, Joonrak Kim and Bongju Jeong *

Department of Industrial Engineering, Yonsei University, 50 Yonsei-ro Seodaemun-gu, Seoul 03722, Korea; sdmworld@yonsei.ac.kr (D.S.); joonrak@yonsei.ac.kr (J.K.)
* Correspondence: bongju@yonsei.ac.kr; Tel.: +82-2-2123-4013

Received: 15 April 2019; Accepted: 29 April 2019; Published: 1 May 2019

Abstract: Globally, many countries are experiencing economic growth while concurrently increasing their energy consumption. Several have begun to consider a low-carbon energy mix to mitigate the environmental impacts caused by increased fossil fuel consumption. In terms of maximizing profits, however, power producers are not sufficiently motivated to expand capacity due to high costs. Thus, the Korean government initiated the Renewable Portfolio Standard (RPS), an obligation to generate a certain proportion of a producer's total generation using renewable energy for power producers with capacities of 500 MW or more, and the Emissions Trading Scheme (ETS), designed to attain a carbon emissions reduction goal. We propose a mathematical model to derive the optimal operational strategy for maximizing power producer profits with a capacity expansion plan that meets both regulations. As such, the main purpose of this study was to obtain the optimal operational strategy for each obligatory power producer. To that end, we defined a 2×2 matrix to classify their types and to conduct scenario-based analyses to assess the impact of major factor changes on solutions for each type of power producer. Finally, for the power generation industry to operate in a sustainable and eco-friendly manner, we extracted policy implications that the Korean government could consider for each type of power producer.

Keywords: renewable portfolio standards; emissions trading scheme; optimal operational strategy; generation expansion plan; energy planning

---

1. Introduction

Globally, interest in future-oriented values such as sustainability or green growth is rising steadily [1,2]. Many countries experiencing economic growth have an accompanying increase in energy consumption [3]. In recent decades, these countries have been considering environmental threats and the risks of depletion of traditional energy sources.

As part of this effort, a new climate regime was launched under the United Nations Framework Convention on Climate Change (UNFCCC) in 2015. At this meeting, each country voluntarily established greenhouse gas (GHG) reduction targets, and adjusted policies are being established to achieve these targets [4]. The main policies include the Renewable Portfolio Standard (RPS) and the Emissions Trading Scheme (ETS), which reduces fossil fuel consumption and carbon emissions.

Korea is the eighth largest energy and electricity consuming country in the world [5,6], as its main industries are energy-intensive, including steel and semiconductor manufacturing. The energy consumption of Korea's main industries is increasing continuously, and electricity consumption represents a large portion of this. However, the Korean generation industry's dependence on highly non-renewable energy (NRE) sources has been an issue. To solve this problem, in 2012, the government implemented RPS, which obligates power producers with 500 MW or more of NRE generation capacity to produce a certain percentage of the previous year's NRE generation capacity from renewable energy (RE) sources. The government aims to have more than 10% of electric power generation come from RE sources by 2023.

At the 21st Conference of Parties, Korea, whose GHG emissions were 7th among Organization for Economic Co-operation and Development (OECD) countries in 2016, declared a national reduction target of 37% compared to business as usual (BAU) by 2030 [4]. In order to achieve this, in 2015, the government started ETS, which assigns companies Carbon Emissions Reductions (CER) based on annual and sectoral GHG reduction targets. Each company complies with this regulation by buying CER surplus, selling CER shortage, or paying penalties for excess emissions.

Korea's major producers in the power generation industry, which accounts for more than one-third of national GHG emissions [7], must follow both regulations. RPS and ETS act as means to implement international commitments, from a national viewpoint, and to create a new paradigm that can affect operations, from an enterprise perspective. Therefore, obligatory power producers (hereafter, power producers) should be aware of the mechanisms of both regulations. They should also provide a stable power supply and establish an efficient future operational strategy.

Korea's power industry is moving towards increasing the private-sector share from the nation-led sector. Thus, from a corporate management perspective, RPS and ETS will affect revenues and costs. Most previous studies have derived energy plans that expand generation capacity and minimize the total cost from a national perspective [1,3,5,8–11]. Therefore, our study aimed at maximizing the profit of power producers that are required to meet electricity demand while operating under both regulations. As such, we developed a mathematical model to maximize profits and provide optimal operational strategies, including capacity expansion plans and RPS/ETS implementation plans, for each power producer. In addition, we proposed a 2 × 2 matrix to classify them for analysis from the power producer's perspective and to analyze the optimal operational strategy for each scenario according to type using this matrix.

The rest of the paper is organized as follows. In Section 2, we review previous studies on both regulations in other countries and on quantitative studies on generation expansion planning (GEP) optimization. Differences between our study and previous studies are highlighted here. Section 3 examines the RPS and ETS mechanisms in Korea in detail and derives an operational scheme oriented towards the power producers regulated by both systems. Mathematical models for deciding operational strategies are developed in Section 4. In Section 5, a scenario-based analysis is conducted to assess how major factor changes affect solutions for power producers in each scenario. Finally, meaningful insights and political implications are provided in Section 6.

2. Literature Review

2.1. Associated with Government Policy

The expansion of RE sources is necessary for sustainable generation in the future. However, the insufficient incentive to expand the RE capacities for a power producer is a problem because it requires a lot of capital for new capacity expansion. Thus, the Korean government implemented RPS and ETS in order to reduce carbon emissions by increasing the proportion of RE power generation.

2.1.1. Renewable Portfolio Standards (RPS)

Many studies on RPS have been conducted including studies on the effects and contributions of an RPS introduction, policy proposals for effective RPS implementation, and portfolio design that reflects RPS uncertainty.

First, some researchers conducted studies analyzing the effect of introducing RPS in the United States (US) [12,13]. The study in Reference [14] found that the introduction of RPS contributed 4.2% and 6.1% to the adoption of wind and solar power, respectively, as a result of excluding political and economic factors. In addition, References [15,16] evaluated the introduction of RPS in consideration of new variables.

Some studies on implementing RPS have been conducted in Korea. First, Reference [17] conducted a preliminary evaluation of Korea's RPS introduction by comparing Feed-in tariffs (FIT) and RPS in terms of capacity expansion, technological progress, cost-effectiveness, and market risk. Furthermore, References [4,18] investigated the current RPS policy in Korea, identified problems, and derived a future outlook based on quantitative indicators. Reference [18] conducted a virtual evaluation of current RPS policy in Korea based on a bottom-up model. Reference [4] analyzed an optimal portfolio for 2050 in the Korean power sector.

Some studies addressed the implementation of RPS in the European Union (EU). The study in Reference [19] proposed an optimization model to increase power supply stability and sustainability. It analyzed the EU policy to compare the benefits of RE sources to traditional energy sources from a cost, risk, and pollutant emissions perspective. In order to minimize costs and risks, the study analyzed the EU RPS status by year and derived the indicators, such as pollutant emissions and target achievement rates. In addition, Reference [20] conducted a study to evaluate the policy coordination impact when implementing RPS and carbon cap-and-trade according to the effective policy mix interval.

2.1.2. Emission Trading Scheme (ETS)

Existing studies on ETS derive a framework for ETS adoption, evaluate the effectiveness of specific state and country ETS policies, and evaluate the impact of ETS on social surplus.

Some researchers have been conducted on ETS framework designs and implementation schemes. First, Reference [21] focused on the development of the ETS regulatory framework in Shenzhen, China, including economic aspects and GHG emissions. An overview of key elements and progress of the Shanghai ETS design was presented in Reference [22]. In addition, Reference [23] conducted an overall evaluation of the ETS framework implemented in Korea in 2015 and analyzed the main points of policy design for consistent policy development. Reference [24] analyzed a scenario-based analysis and simulation for establishing a multi-regional integrated ETS scheme using computable general models in China, the US, Europe, Australia, Japan, and Korea.

Many studies regarding implementing ETS schemes have been conducted lately. The study in Reference [25] investigated the problems that arise when implementing an ETS scheme, which considered the decision making of power generation companies in the management areas of local government. References [26,27] studied the price effects of carbon credits.

On the other hand, some studies discussed the risks of introducing ETS and the factors that hinder the cost-effectiveness of ETS. Reference [28] classified risks (market risk, policy risk, green investment risk, etc.) from the supply chain perspective when implementing ETS and derived a strategic portfolio for risk mitigation. In addition, Reference [29] evaluated Korean ETS based on factors that actually limit the cost-effectiveness of ETS and established an operational strategy to control them. Establishing an operating system for ETS that takes into account realistic constraints and risks can be used as a sustainable policy tool.

2.2. Generation Expansion Planning (GEP) Optimization

Generation Expansion Planning (GEP) optimization is a research area that derives the optimal power generation portfolio composition. This research area includes many studies, mainly focused on the development of expansion plans for power generation capacities using mixed-integer programming, and on considering uncertainty in expansion plans. In addition, many studies have analyzed the effects of various factors (incentive systems, climate change, decentralized development, etc.) in the models.

Some researchers have presented a new framework for considering uncertainties in GEP. First, Reference [8] proposed a model based on mixed-integer programming that reflected the uncertainties of electricity demand, investment, and operating costs using robust optimization techniques. Reference [30] derived an expansion plan that included stability and reliability constraints, and Reference [9] presented a GEP model that considered input variables uncertainty.

References [31,32] studied a power plant expansion problem using mixed-integer programming models. Reference [32] included both supply-side and demand-side approaches. The study in Reference [33] solved the problem by considering three objective functions: project lifetime economic return, minimization of CO_2 emissions, and minimization of fuel price uncertainty. Reference [34] presented a multistage expansion planning problem of a distribution system that considered both distribution networks and distributed generation. That model was based on cost minimization considering investment cost, maintenance cost, production cost, loss cost, and residual energy cost, and the energy loss cost reflected by piecewise linear approximation. Reference [35] studied generation expansion planning to expand the share of RE. This study developed a multi-objective (minimizing the total cost, maximizing generation at the peak load, and maximizing the contribution of RE) optimization model and analyzed the Brazilian case based on a scenario.

Other researchers have presented two-step decision-making methods. Reference [36] addressed the maximization of investor returns at the upper level and the maximization of social surplus at the lower level. In addition, Reference [37] suggested heuristic-based dynamic programming for expansion plans. The study in Reference [10] defined the GEP problem of a restructured power system reflecting multi-period uncertainty for facility investment decisions in terms of price maker and suggested a framework accordingly. Reference [1] presented a power generation expansion (PGE) model incorporating various low-carbon elements for low-carbon economic growth. For the scenario-based experiments considering a low carbon situation, a compromised modeling approach was used to reduce model complexity.

A study on the optimum operational strategy for a power producer considering the sales and purchasing of renewable energy certificates (REC)/CER, penalty cost, and capacity investment cost under the simultaneous regulation of the RPS and ETS systems has not yet been conducted. Specifically, most quantitative studies considering RPS or ETS were conducted from a country perspective (cost minimize) [38]. However, this study proposes a mathematical model for maximizing power producer profits. This is because the viewpoint of the study is likely to be more effective in achieving power producer sustainability, given that the role of private power producers is expanding.

3. An Operational Scheme Considering a Combined Regulatory Environment

This section examines the RPS and ETS mechanisms in Korea's power generation industry. It defines an integrated operational scheme for power producers under both regulations.

- Renewable Portfolio Standards (RPS)

Sixteen countries around the world implemented the RPS [39], and the Korean government made it mandatory for power producers with 500MW or more of generation capacity (18 companies in 2017) to supply a certain proportion of RE. Table 1 shows the yearly RPS target ratio in Korea's power generation industry.

Table 1. RPS target ratio. (RPS: Renewable Portfolio Standard).

Year	2017	2018	2019	2020	2021	2022	2023 on
RPS target ratio (%)	4	5	6	7	8	9	10

The government assigned the REC targets, calculated by multiplying the amount of electric power from NRE in the previous year and the yearly target ratio to power producers. REC targets are satisfied in three ways: self-generation, purchase of RECs in the market, or penalty payment. Figure 1 is a diagram of the RPS mechanism with the power producer in the center.

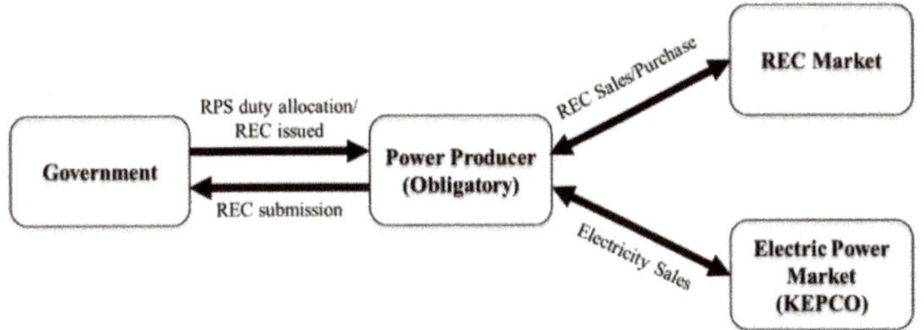

Figure 1. Diagram of the RPS mechanism. (KEPCO: Korea Electric Power Corporation).

As shown in Figure 1, power producers need to satisfy their obligations either by electric power generation using RE or by purchasing RECs in the market. To deal with the RECs themselves, the addition of RE capacity is necessary. However, since the investment cost for each RE source differs, the issue is that, if the same weight is given, the investment of power producers in a certain RE source can be biased. In order to prevent such biases, the Korean government has set REC weights differentiated for each energy source. Therefore, each power producer has to consider different weights when deciding how much of an energy source to add and which RECs to buy and sell in the REC market.

- Emissions Trading Scheme (ETS)

The Korean government has set a carbon emissions reduction target of 37% versus BAU by 2030, and it began implementing the ETS in January 2015. Because reduction goal in the power generation sector is very high according to the National Emissions Allocation Plan [7], the government assign the annual CERs to the power producers Figure 2 diagrams the ETS operation mechanism for power producers reflecting the situation described below.

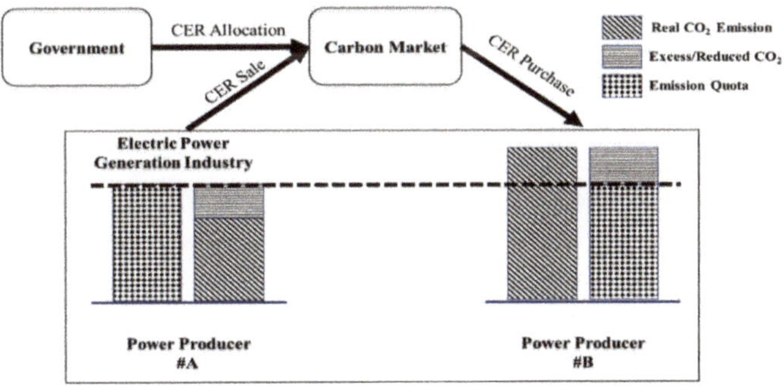

Figure 2. Diagram of the ETS mechanism. (ETS: Emissions Trading Scheme).

- Operational Scheme Considering a Combined Regulatory Environment

While most countries that implement RPS are obliged to supply RE to electricity sellers, Korea is uniquely obliged to supply each power producer because the government monopolizes the electricity market. Korea's power producers must comply with RPS duties and ETS emissions quotas. Therefore, a decision-making system that maximizes profits by considering REC/CER price changes, different power generation and operating costs for each energy source, new capacity investment costs, periodic obligations, and emission quotas is necessary. This study designed an operational scheme for Korea's power producers regulated by RPS and ETS simultaneously, shown in Figure 3.

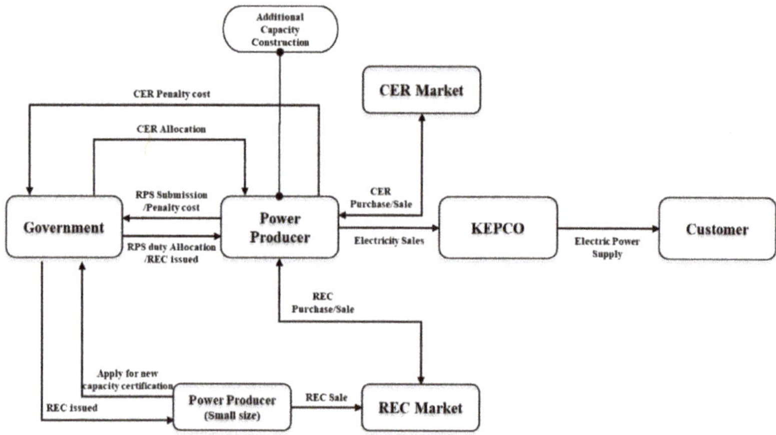

Figure 3. Integrated operational scheme for Korean power producers.

- Input Parameter Estimation for the Mathematical Model

For use as input values in the mathematical model, we first must estimate SMP, REC, and CER prices that change in the market.

First, this study uses the SMP estimation method in the previous study [40]. For REC and CER, forecasting is somewhat difficult due to the lack of existing data because the market operation period is short. Therefore, we use the average change rates of REC/CER to predict each price for the planning horizon to simplify parameter estimation.

Additionally, the total number of power producers considered in this study is 18 companies in Korea, which accounts for about 90% of Korea's total power generation capacity. This proportion (θ) will be a parameter to adjust the annual demand and emissions quota. For example, if the forecasting demand for 2018 is 100 TWh, the demand for power producers will be 90 TWh. These values are used as input values in our model.

4. Mathematical Modeling

4.1. Problem Structure and Definition

This study proposes a mathematical model based on mixed-integer linear programming to derive optimal operation strategies for Korea's power producers to maximize profit. Section 3 addressed the integrated operational scheme. The operational strategy corresponds to the value of the arrows derived from the power producers, composed of the decision variables of the proposed mathematical model. The decision variables are $X_{e,m,t}$, denotes the new generation capacity completed by each period, $x_{e,m,t,k}$, constructs the generation capacity of each period to achieve, and $X_{e,m,t}$ and $Y_{e,m,t,k}$, the binary variables for establishing the power generation plan. To calculate various revenue/cost values, the model uses the sale/purchase/non-fulfillment of REC and CER as variables. Figure 4 shows the structure of the problem proposed in this study. The blue variables have a positive effect on profit, while the red variables have a negative effect on profit.

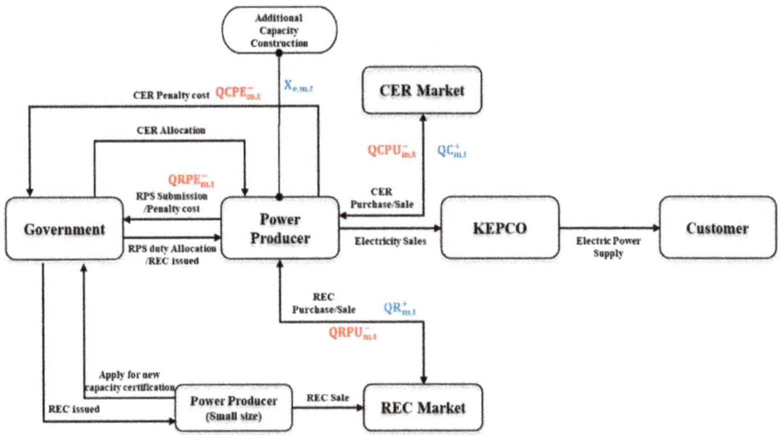

Figure 4. Problem structure.

4.2. Mathematical Model for Operational Strategy

This study uses the electricity demand and reserve margin of each planning period, based on data from the 7th basic plan for long-term electricity supply and demand [41]. This study considers 18 power producers (called M1 to M18) with capacities of 500 MW or more, which is the RPS standard. In addition, the capacity expiration and expansion plans included in the 7th basic plan are known, and power producers can sell their generated electric power to KEPCO, and the extra REC and CER can be sold to the market without limitation. However, this study assumes that the amount of additional REC/CER purchases are different depending on the size of the power producer. The annual unit generating and operating costs are known for all planning periods for each generation source [3,42–46]. Lastly, power plant expiration has not been confirmed in the 7th basic plan for the whole period, and the lifetimes of newly added power plants are not considered in this study.

4.2.1. Objective Function

The purpose of this model is to maximize the total profit of Korea's power producers. Power producers generate electricity to meet annual demand, sell it to KEPCO, and earn revenue by selling additional REC and CER. In terms of cost, additional costs for purchasing REC and CER, penalty costs, power generation and operation costs, and investment costs for new capacity expansion are taken into account. The objective function is expressed as Equation (1).

Minimize Total Profit (TP):

$$TP = \left\{ \sum_{m,t} \left(ES_{m,t} + QR_{m,t}^{+}PR_t + QC_{m,t}^{+}PC_t \right) \right\}$$
$$- \left\{ \sum_{m,t} \left(QRPU_{m,t}^{-}PR_t + QCPU_{m,t}^{-}PC_t + REC_{m,t}^{pen} + CER_{m,t}^{pen} + GOC_{m,t} + Inv_{m,t} \right) \right\}. \tag{1}$$

To simplify the objective function, the model used the expressions such as $ES_{m,t}$, $REC_{m,t}^{pen}$, $GOC_{m,t}$, $Inv_{m,t}$ and calculated as follows:

$$ES_{m,t} = \varepsilon \sum_e Hr_e Eff_e Capa_{e,\,m,t} PE_t, \forall m, t \tag{2}$$

$$REC_{m,t}^{pen} = 1.5PR_t \times max\left(0, QRPE_{m,t}^{-}\right), \forall m, t \tag{3}$$

$$CER_{m,t}^{pen} = 3PC_t \times max\left(0, QCPE_{m,t}^{-}\right), \forall m, t \tag{4}$$

$$GOC_{m,t} = \varepsilon \sum_e \left(Hr_e Eff_e Capa_{e,\,m,t} genc_{e,t} \right), \forall m, t \tag{5}$$

$$\text{Inv}_{m,t} = \sum_e \sum_{k=1}^{t-1} (1+\gamma)^{-k}(\text{IC}_e x_{e,m,t,k}), \forall m, t. \tag{6}$$

Equation (2) is the electricity sales of each power producer in period t, and Equations (3) and (4) are the penalty costs due to failing to achieve REC and CER targets. Equation (5) is the generation and operation cost of each power producer in period t, and Equation (6) is the investment cost for the expansion capacity.

4.2.2. Constraints

This chapter considers the following constraints.

$$\text{Capa}_{e,\,m,t} = \text{Capa}_{e,\,m,t-1} + X_{e,m,t} - \exp_{e,m,t-1} + \text{add}_{e,m,t}, \forall e, m, t \tag{7}$$

$$\varepsilon \left(\sum_{e,m} \text{Hr}_e \text{Eff}_e \text{Capa}_{e,\,m,t} \right) \geq \theta(1 + \text{RM}_t)D_t, \forall t \tag{8}$$

$$x_{e,m,t,k} \leq \delta_e Y_{e,m,t,k}, \forall e, m, t, k \tag{9}$$

$$X_{e,m,t} = \sum_{k=1}^{t-1} x_{e,m,t,k}, \forall e, m, t \tag{10}$$

$$\sum_m X_{e,m,t} \leq \text{Capa}^P_{i \in e,t}, \forall i \in e, t \tag{11}$$

$$\text{Inv}_{m,t} \leq \text{Budget}_{m,t}, \forall m, t \tag{12}$$

$$\text{ETS}^{\text{Quota}}_{m,t} + \text{QC}^-_{m,t} = \sum_e \left(\text{Hr}_e \text{Eff}_e \text{Capa}_{e,\,m,t} \text{CF}_e \right) + \text{QC}^+_{m,t}, \forall m, t \tag{13}$$

$$\text{QCPE}^-_{m,t} + \text{QCPU}^-_{m,t} = \text{QC}^-_{m,t}, \forall m, t \tag{14}$$

$$\text{QCPU}^-_{m,t} \leq \text{CUL}_m, \forall m, t \tag{15}$$

$$\sum_{e,m} \left(\text{Hr}_e \text{Eff}_e \text{Capa}_{e,\,m,T} \text{CF}_e \right) \leq \sum_m \text{ETS}^{\text{Quota}}_{m,t}, \text{ where } t = T \tag{16}$$

$$\text{RPS}^{\text{Target}}_{m,t} \times \varepsilon \sum_{j \in e} \left(\text{Hr}_j \text{Eff}_j \text{Capa}_{j,\,m,t-1} \right) + \text{QR}^+_{m,t} = \varepsilon \sum_{i \in e} \left(\text{Hr}_i \text{Eff}_i \text{Capa}_{i,\,m,t} \text{WR}_{i \in e} \right) + \text{QR}^-_{m,t}, \forall m, t \tag{17}$$

$$\text{QRPE}^-_{m,t} + \text{QRPU}^-_{m,t} = \text{QR}^-_{m,t}, \forall m, t \tag{18}$$

$$\text{QRPU}^-_{m,t} \leq \text{RUL}_m, \forall m, t \tag{19}$$

$$\sum_{i \in e,m} \left(\text{Hr}_i \text{Eff}_i \text{Capa}_{i,\,m,t} \times \text{WR}_{i \in e} \right) \leq \sum_m \left(\text{RPS}^{\text{Target}}_{m,t} \times \sum_{j \in e} \left(\text{Hr}_j \text{Eff}_j \text{Capa}_{j,\,m,t-1} \right) \right), \text{ where } t = T \tag{20}$$

$$\text{All variables are non} - \text{negative}, \forall e, m, t, k. \tag{21}$$

Equation (7) represents the cumulative capacity updating formula for each energy source by power producer and period, including new capacity to be completed during the period, planned capacity increases during the period, and expired capacity in the previous period. Equation (8) is a constraint for meeting demand, and Equation (9) represents the feasible expansion amount that can be added in period k. Equation (10) expresses that the sum, up to the amount of (t − 1) period added by period k, is completed in t period. Equation (11) is used to set the RE potential capacity upper limit, and Equation (12) is an expression that shows the investment budget constraints for each company.

Equations (13) through (16) reflect the ETS policy. Equation (13) is the CER balance equation for each power producer. $\text{QC}^-_{m,t}$ denotes the CER shortage and expresses the sum of additional purchase amounts and the penalty amounts from Equation (14). However, Equation (15) represents an upper limit on the number of additional purchases. Equation (16) expresses constraints on achieving the goals set by the country in the final year of the planning horizon.

Equations (17) through (20) reflect the RPS policy. Equation (17) is the REC balance equation for each power producer. $QR^-_{m,t}$ used in this equation means that REC shortage and can be expressed as the sum of the additional purchases and the penalty amounts given by Equation (18). Equation (19) is used to set an additional purchase upper limit. Equation (20) represents the constraints for achieving the goal set by the country in the final year of the planning horizon. Finally, Equation (21) implies that all variables used in this model should be non-negative.

Section 5 will perform scenario-based analyses, and the mathematical models applied to all scenarios are basically identical. However, this study can discover some differences in input parameters for each factor considered in each scenario, and the constraints may require changes. For example, consider Equations (22) and (23) below.

$$\begin{cases} X_{e,m,t} = 0, & \text{where } e = 10 \text{ and } m \neq 1, \; \forall t \\ X_{e,m,t} = 0, & \text{where } e = 8 \text{ or } e = 10, \; \forall m, t \end{cases}$$

Equation (22) is a constraint that makes it impossible for other companies to expand nuclear power plants, as nuclear power plant expansion in Korea is only possible for M1. Equation (23) is a restriction that makes it impossible to add new coal-fired and nuclear power plants.

5. Scenario-Based Analysis

This study examines how various factor changes affect the optimal operational strategy of power producers using the proposed model. First, we define four scenarios, and basic information for each scenario is shown in Table 2. In summary, Scenario 1 aims at creating a reference case by deriving operational strategy based on the current situation in Korea as an input value. Scenario 2 analyzes the impact of strengthening RPS duty ratio through nuclear and coal phase-out, and Scenario 3 aims to analyze the impact of tightening the carbon emissions regulations by increasing CER prices. Lastly, Scenario 4 is to analyze the impact of reaching the RE grid parity.

Table 2. Basic information for each scenario.

No	Description	Control Factor			
		Enhanced RPS Duty	Nuclear Option	CER Price	Generation Cost
SC1	Reference case	-	-	-	-
SC2	Nuclear and coal phase-out scenario: Considering strengthened RPS duty ratio	O	O	-	-
SC3	Impact of increasing the unit CER prices	-	-	O	-
SC4	Influence of reaching grid parity: Changing the generation and operation cost of RE sources	-	-	-	O

5.1. Input Data

In this model, it is assumed that the electricity demand and reserve margin in the entire planning horizon are already known, beginning in 2017 [41]. However, the demand is scaled using the initial generation capacity ratio of power producers under both regulations in Korea. (See Table 3)

The initial generation capacity data for each power producer is collected from the total generation capacity report. It is assumed a 3.59% transmission and distribution loss factor, constant during the entire period [43]. In addition, planned expanded and expired generation capacity amounts for each power producer were gathered from the 7th basic plan [41], and the realizable potential of each RE source was obtained from previous research and statistics [42,47].

Table 3. Demand (TWh) and reserve margin (%).

Year	2017	2018	2019	2020	2021	2022	2023	2024	2025	2026	2027	2028	2029
Demand (TWh)	498	522	544	563	580	597	612	627	642	657	671	684	698
Reserve Margin (%)	26.3	24.9	23.7	23.2	26.8	27.7	25.3	22.4	21.2	21.3	21.4	21.5	21.6

Next, the annual investment budget of each power producer during the planning horizon is deterministic. In this case, it is assumed that this value starts from 7% of initial sales and increases by an annual average of 3%. Data such as the capacity factor for each energy source, the investment cost per unit capacity addition, carbon emissions coefficients, generation and operation costs, and historical price data to use for estimation were collected from previous research [3,6,42,44–49] and the Electric Power Statistics Information System (EPSIS) (see Table 4) [43]. In scenario 4, considering the attainment of grid parity, it is assumed that the generation and operation costs of RE decrease linearly, lower than that of the lowest among NRE after 2021.

Table 4. Investment, capacity factor, generation and operation costs, and CO_2 emissions coefficient.

Type	Energy Source	Investment Cost (M KRW/MW)	Capacity Factor (h)	Generation and Operation Costs (KRW/MWh)		CO_2 Emissions Coefficient (kgCO$_2$/MWh)
				Low	High	
Renewable Energy (RE)	Solar PV	3500	1314	65,000	271,000	33
	Wind	2500	2891	35,000	83,000	12
	Hydro	1170	4643	32,000	108,000	4
	Biomass	3500	7271	83,000	119,000	18
	Fuel cell	5680	7446	115,000	181,000	221
	Ocean	3824	1752	249,000	271,000	37
Non-renewable Energy (NRE)	LNG	620	6132	54,000	87,000	450
	Coal	540	7446	70,000	162,000	950
	Oil	1908	7621	114,000	129,000	680
	Nuclear	1562	7884	60,000	147,000	16

Lastly, the annual RPS target ratio uses the values specified in the regulations. Scenario 2 assumes a planned annual target ratio that can achieve 20% in the last period to reflect the government's intent to increase the proportion of RE through the implementation of a nuclear and coal phase-out roadmap (See Table 5).

Table 5. Strengthened RPS target ratio plan.

Year	2017	2018	2019	2020	2021	2022	2023	2024	2025	2026	2027	2028	2029
RPS Target Ratio (%)	4.0	6.0	8.0	10.0	12.0	13.0	14.0	15.0	16.0	17.0	18.0	19.0	20.0

Table 6 shows the annual carbon emissions targets for the power generation industry, obtained by applying a linear interpolation method to the initial and final year target values.

Table 6. Total CO_2 emissions targets.

Year	2017	2018	2019	2020	2021	2022	2023	2024	2025	2026	2027	2028	2029
CO_2 Target (MCO$_2$ton)	247	249	250	252	254	255	257	259	260	262	264	265	267

5.2. Results

5.2.1. Scenario 1: Reference Case

Figure 5 presents the amount of newly installed capacity of each energy source over the planning horizon. First, it seems that biomass generation capacity steadily expands, though the investment cost is high. This is because biomass generation is an attractive means to fulfill the RPS obligations, in that it has a higher capacity factor than other RE sources and can receive a REC. Second, wind and fuel cell generation tend to expand more after 2023, after most power producers have achieved their obligations. This preference increase is to improve profits by adding more energy sources with higher REC weights after eliminating the RPS penalty from not satisfying the RPS regulation conditions. Lastly, hydropower generation consistently increases. This happens because it has advantages such as low generation and operation costs, low carbon emissions, and RECs can be issued in the case of small-hydro generation.

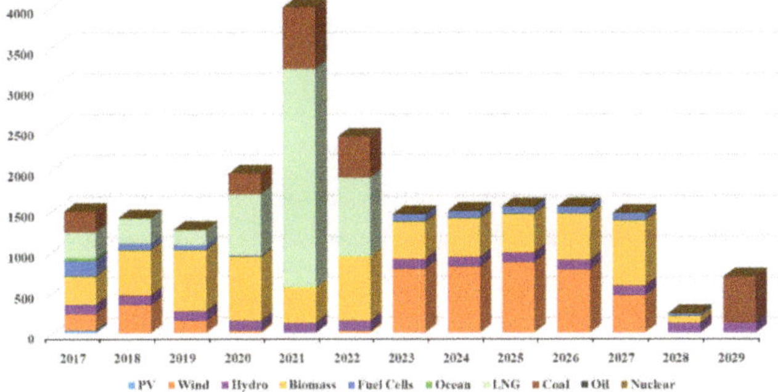

Figure 5. Newly installed capacity of each energy source in SC1.

Figure 6b presents the changes in capacity and generation share of RE/NRE from 2016 to 2029, respectively. In 2029, 12.6% of the power producer's total installed capacity will be comprised of RE sources, accounting for 9.7% of generation. In addition, the rates of capacity and generation increase from 2016 to 2029 show that RE increases more rapidly than NRE. As such, the government's regulations (RPS and ETS) have an effective impact on the expansion of RE proportion in the power producer's energy mix.

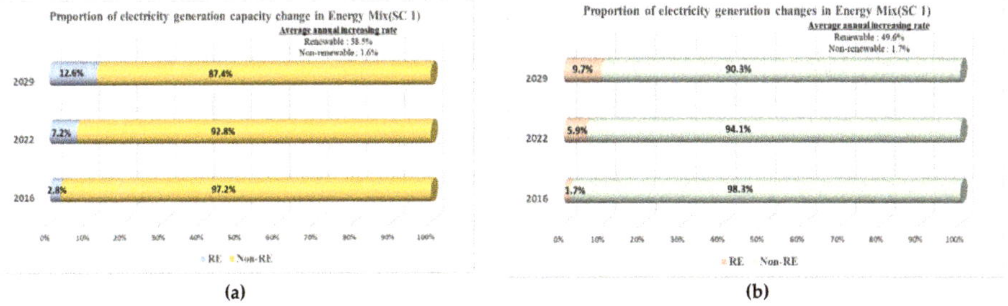

Figure 6. The proportion of RE/NRE capacity and generation. (**a**) Trends in the proportion of RE/NRE capacity (SC1); (**b**) trends in the proportion of RE/NRE generation (SC1). (RE: Renewable Energy, NRE: Non-renewable Energy).

The main purpose of this study was to derive an optimal operational strategy for each power producer. In order to analyze from each individual power producer's perspective, we created a power producer classification matrix. This matrix classified power producers into four types, with initial RE capacity proportion and initial sales as two axes. Clockwise from the right top in Figure 7, the quadrants are defined as industry leaders (IL), reliable supporters (RS), industry followers (IF), and aggressive participants (AP). Characteristics of each segment are described in Table 7.

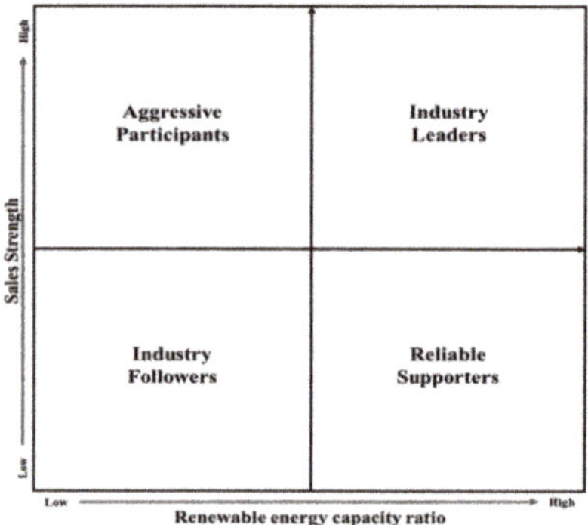

Figure 7. Power producer type matrix.

Table 7. Characteristics of each segment.

Segment	Characteristics
Industry Leaders	Power producers whose initial RE capacity proportion and initial sales are both high The group with the largest installed capacity share in the entire group
Reliable Supporters	Although initial RE capacity retention ratios are high, power generation is less than that of IL, and with lesser sales Power producers with a reliable response to government regulation
Industry Followers	The group with the largest number of producers Relatively low initial sales and initial RE capacity share Power producers who accommodate requests slowly to minimize the damage caused by government regulations
Aggressive Participants	The group with higher sales but low RE capacity retention ratio Power producers with a motive force to maximize profit through aggressive additional RE investments

In order to analyze the power producer's behavior in each segment, we first divided 18 power producers into four categories and analyzed the change in RPS duty achievement. First, since the IL segment has a large amount of installed capacity, it takes time to achieve RE duty. Therefore, instead of investing heavily in the early period, it also maximizes profits by combining appropriate long-term investment and profitable activities, such as selling electricity and excess CER.

The RS segment has a smaller generation capacity than the IL segment, and the RE capacity retention ratio is high, so they easily satisfy the RPS duty. Therefore, to minimize the penalty expenditure, they achieve the RPS duty from the beginning to maximize profits.

Finally, power producers in the IF segment usually maintain proper investment from the beginning, but it takes a while to achieve the RPS duty because of the low initial RE capacity retention ratio compared to the RS segment. Figure 8 shows the change of RPS duty ratio among 11 power producers in the IF segment. Unlike other power producers, M10 and M16 show a high achievement ratio from the beginning.

This is because M10 and M16 have larger sales per unit generation capacity (since they provide city gas or solar operations and maintenance business in addition to electric power generation) than other power producers in the same IF segment. This helped them pursue profit maximization through aggressive RE investments. Thus, it can be inferred that power producers with large sales per unit generation capacity (like M10 and M16) will be effective in maximizing profits by following the strategy corresponding to the AP segment because the same investment has a greater impact on the energy mix.

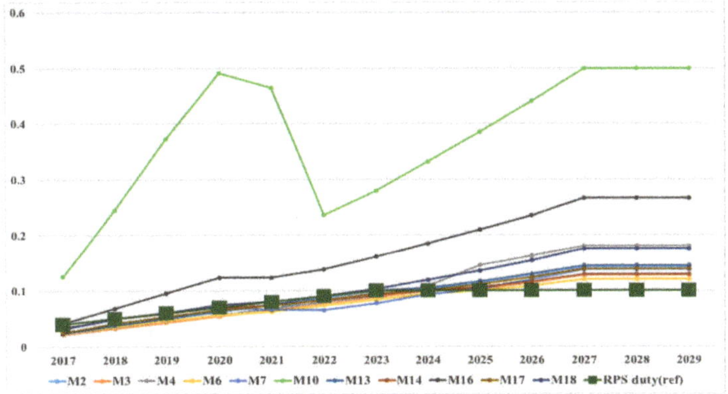

Figure 8. RPS ratio changes of each power producer in the IF segment.

A growing number of power producers will be regulated by RPS. Some of them will be able to make aggressive investments due to large sales, although the RE capacity ratio is low, and they are clearly distinguishable from the IF segment. Therefore, in order to consider such characteristics, it is more realistic and suitable to divide the group into four segments, as shown in Figure 9.

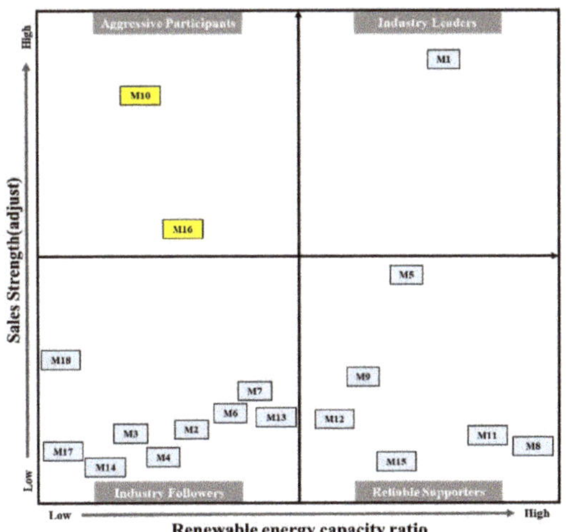

Figure 9. Power producer type matrix (adjusted).

5.2.2. Scenario 2: Nuclear and Coal Phase-out Scenario Considering a Strengthened RPS Duty Ratio

Scenario 2 considers the abolition of 6 nuclear power plants (totaling 8.8 GW) and 10 coal-fired power plants (totaling 3.34 GW), the prohibition on the addition of new nuclear and coal-fired power plants, and the strengthened RPS duty ratio as a constraint.

Compared with the reference scenario, this result shows the capacity of all RE sources increases more rapidly (see Figure 10). In particular, fuel cells with a higher REC weight are installed to meet the RPS duty ratio early on. In addition, wind and biomass generation capacity rise more than in the reference scenario. This indicates that RE sources, which can generate revenue through REC and can minimize penalties from excess carbon emissions, can play an alternative in a situation where nuclear power generation, which accounts for a very large proportion of the national energy mix and has very low carbon emissions, must be replaced. However, to meet the demand in the nuclear and coal phase-out environment, there is a limit to the addition of solely RE sources. Thus, the new installation of competitive liquefied natural gas (LNG) generation capacities among NRE sources tends to increase quickly.

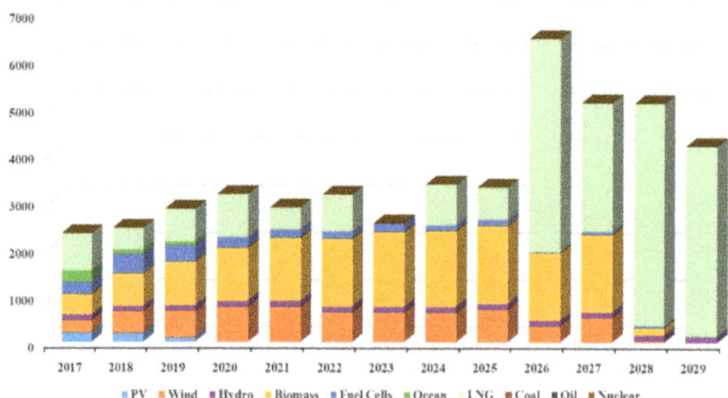

Figure 10. Newly installed capacity of each energy source in SC2.

Figure 11a,b shows that 20.8% of the total power producers generation capacity and 17.8% in terms of generation amount will be composed of RE sources in 2029 (see Figure 12). In addition, we increased the investment budget proportion by 15%, and most power producers tend to aggressively increase RE capacities to gain more REC sales. This tendency is most noticeable in the RS and AP segments. Furthermore, power producers with relatively large capacity in the IF segment continue to make better decisions between payment of penalties and additional investment in achieving their duties. In conclusion, in order to achieve the strengthened duty ratio, the government can use a policy that causes the IF segment (with the most power producers) to be more burdened with penalty payments. Additionally, the RS and AP segments, which are reliably responsive to government regulations, may apply supporting policies, such as loan interest benefits on investment expenditure.

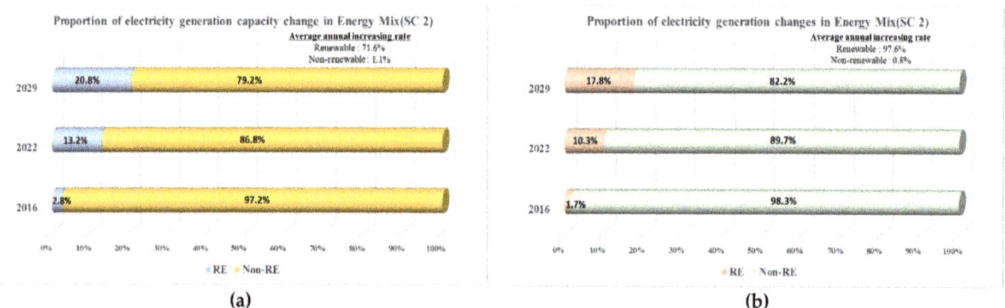

Figure 11. The proportion of RE/NRE capacity and generation. (**a**) Trends in the proportion of RE/NRE capacity (SC2); (**b**) trends in the proportion of RE/NRE generation (SC2).

5.2.3. Scenario 3: Impact of Increasing the Unit CER Prices

The CER price of Korea set in the reference scenario is somewhat low, and thus we considered this scenario with increasing CER price conditions (annual averages of 5%, 10%, and 15%) to more strongly regulate emissions. Ultimately, we examined how the power generation industry and power producers from each segment would react to such CER price increases.

First, the changes in RE/NRE generation amounts according to the CER price increases are shown in Figure 13.

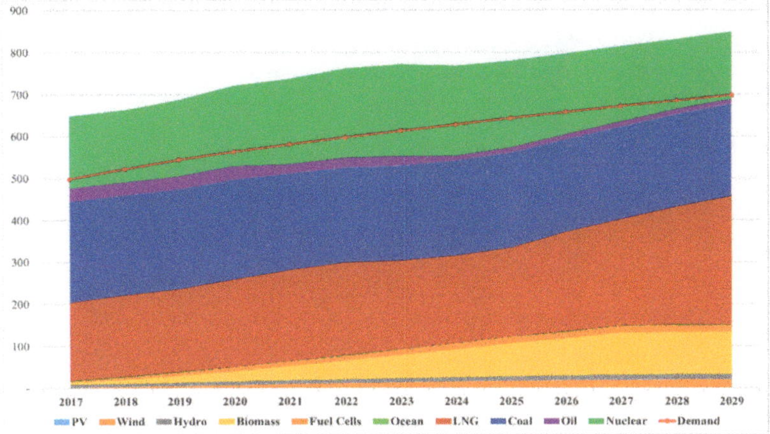

Figure 12. Power generation changes in SC 2 (TWh).

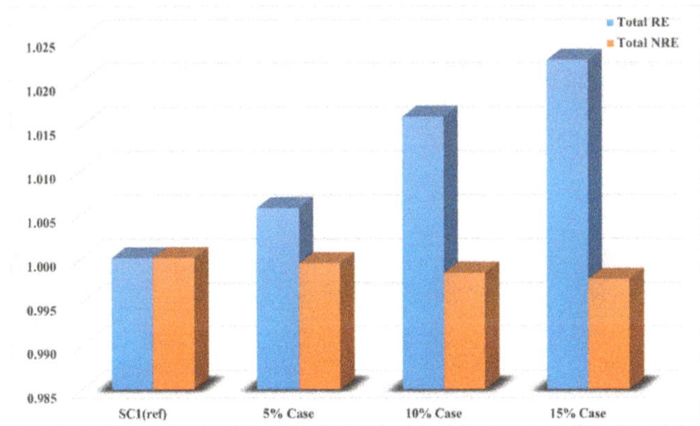

Figure 13. Power generation changes of RE/NRE considering CER price increases.

It shows the ratios of RE/NRE generation amounts by CER price increase level when the generation amount of the reference scenario is set to 1 in 2029. As the annual average increase rate rises, the total generation amount is the same in 2029, but the RE generation amount increases and the NRE generation amount tends to decrease. For instance, at an annual average CER price increase of 15%, the RE generation increases by 1800 GWh compared to the reference scenario, close to the amount of solar generation capacity (1400 MW). However, the generation capacity represents only a 0.1% difference. This means that power producers with the same energy mix will pursue a different power plant utilization strategy according to the CER price change.

Figure 14 shows the changes in RE/NRE generation changes for each segment according to the CER price change when the reference scenario power generation is set to 1 in 2029. As the price increase rate rises, the NRE power generation tends to decrease. Table 8 shows the actual generation (TWh) with CER price changes according to representative companies in each segment. In summary, since the IL segment accounts for a large share of the total installed capacity, additional NRE generation, the amount that is reduced by the power producers of other segments, tends to happen to satisfy the demands. In addition, the AP segment shows a tendency to increase the RE generation amount most sensitively as the CER price increases. Meanwhile, the RS segment steadily increases RE generation as the CER price increases, for up to a 5% increase, and the power producers of the IF segment do not increase RE generation due to the small penalty, but they start responding at the 10% increase level. Therefore, a more effective approach would be to apply differentiated penalty rates to each segment, divided into more than and less than certain emissions amounts, at current low prices of the CER market.

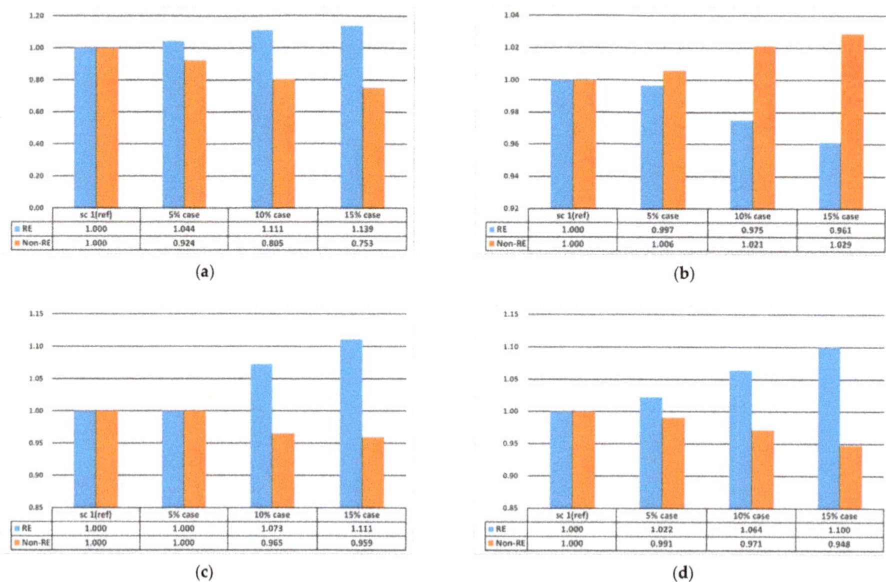

Figure 14. RE/NRE power generation changes with CER price increases in: (**a**) AP segment (M10); (**b**) IL segment (M1); (**c**) IF segment (M18); (**d**) RS segment (M5).

Table 8. RE/NRE power generation (TWh) with CER price changes for each segment.

RE/NRE Generation (TWh)		Scenario 1 (Reference)			5% Case			10% Case			15% Case		
		2017	2023	2029	2017	2023	2029	2017	2023	2029	2017	2023	2029
M1	RE	3.42	11.26	21.53	3.42	10.88	21.46	3.42	9.68	20.99	3.42	9.18	20.69
(IL)	NRE	174.4	262.1	310.7	174.4	264.2	312.6	174.4	269.7	317.3	174.4	272.6	319.7
M5	RE	1.65	5.27	8.38	1.65	5.46	8.56	1.65	5.77	8.92	1.68	6.07	9.21
(RS)	NRE	72.03	75.54	75.54	72.03	74.84	74.84	72.03	73.39	73.39	71.63	71.63	71.63
M10	RE	0.45	5.74	5.74	0.45	3.49	5.99	0.45	3.87	6.38	0.45	4.03	6.54
(AP)	NRE	5.85	15.62	15.62	5.85	14.44	14.44	5.85	12.57	12.57	5.85	11.76	11.76
M18	RE	0.18	1.41	1.41	0.18	0.88	1.41	0.18	0.96	1.52	0.18	0.97	1.57
(IF)	NRE	10.14	10.88	10.88	10.14	10.88	10.88	10.14	10.5	10.5	10.14	10.43	10.43

Next, we examine the emissions intensity trend (tCO_2/MWh) according to the CER price change from the power generation industry and power producer's side.

The IL segment, which accounts for more than 20% of total capacity, is the group with the highest reactivity according to the increase in demand, and the emissions intensities of companies in that group are similar to the increase in demand. The RS segment also shows an increase in emissions intensity versus the same year as CER price increases. What is unusual is that, even if the CER price increase rate rises, the emissions intensity increases. This has a logical explanation: the NRE generation capacity of the IL and RS segments increases to compensate for the decrease in NRE generation caused by the decrease in the AP and IF segment emissions (see Figure 15, Figure 16a–d).

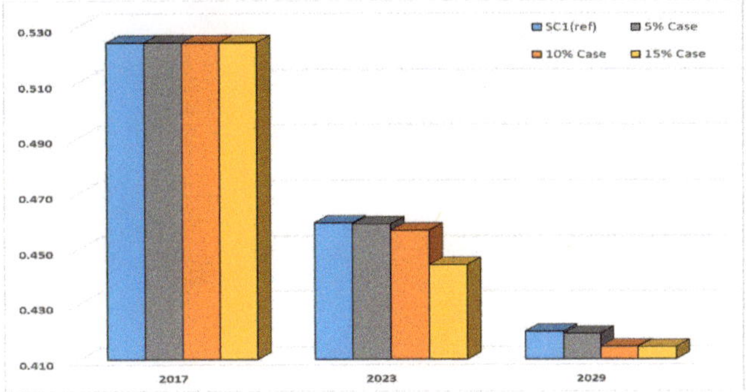

Figure 15. Changes in the emissions intensity (tCO$_2$/MWh) of the power generation industry with CER price increases.

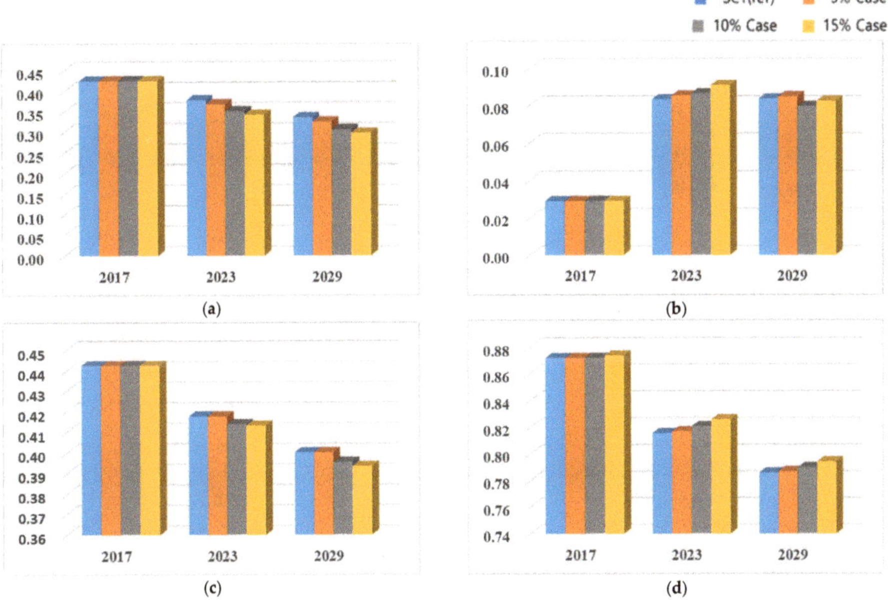

Figure 16. Changes in the emissions intensity (tCO$_2$/MWh) with CER price increases in: (**a**) AP segment (M10); (**b**) IL segment (M1); (**c**) IF segment (M18); (**d**) RS segment (M5).

Next, we examined the change in the RPS duty ratio with CER price changes for each segment. Figure 17 shows the yearly RPS target ratio, and that the level of achievement differs for each segment. The AP segment presents the greatest increase in RPS ratio, and both the IF and RS segments increase steadily beyond the RPS target ratio. On the other hand, the IL segment is close to the target in the latter half of the planning horizon, and even if the CER price increases, it does not reach the goal. This can be understood in the same manner as for the emissions intensity and it is a phenomenon caused by the fact that power producers with a large share in the whole industry have to compensate for the decrease in NRE generation because of the increase in the RPS ratio of the remaining segments. This indicates that the government should not set RPS targets in a lump sum regardless of the power producer's type, but that differentiating the regulatory level based on capacity size and RE/NRE generation will be more effective in achieving the RPS goal.

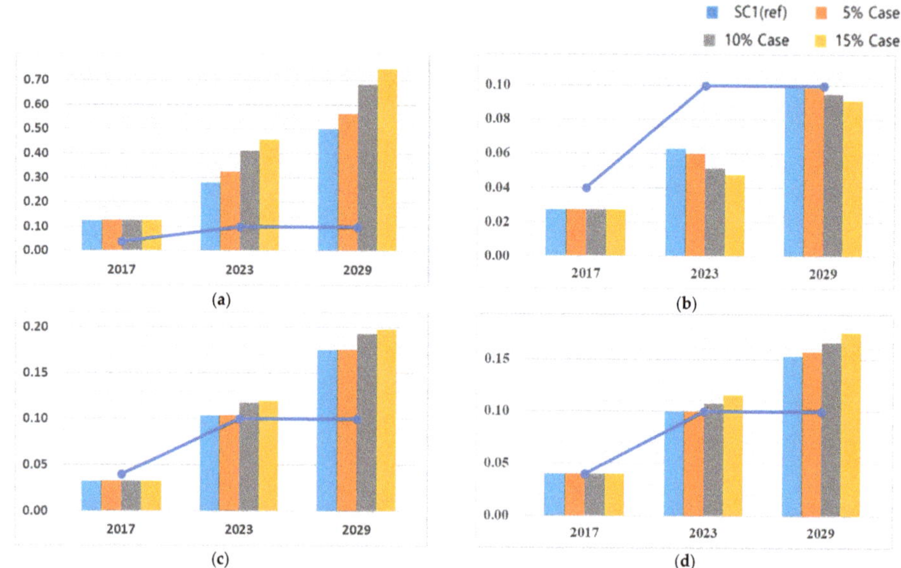

Figure 17. RPS ratio changes with CER price increases in: (**a**) AP segment (M10); (**b**) IL segment (M1); (**c**) IF segment (M18); (**d**) RS segment (M5).

In addition, except for SC2, achieving RPS targets through PV capacity expansion are rarely executed. It is due to the low capacity hours, low power generation efficiency and high generation and operation cost caused by the low amount of sunlight in Korea. Therefore, policy support (increasing power generation subsidies, improving REC weights, etc.) is needed to make the power producers consider solar energy as an attractive generation capacity expansion option in the future.

5.2.4. Scenario 4: The Influence of Reaching Grid Parity by Changing the Generation and Operation Costs of RE Sources

In the early stages of the planning horizon, the generation and operation costs of RE are higher than those of NRE, so there is a tendency to expand coal and LNG, which are both economically superior to other sources, and the RPS duty ratio is achieved primarily by using biomass and small hydropower. After reaching grid parity, profits seem to be created by increasing new capacity investments in fuel cells with high REC weights in order to maximize the cost-benefit of RE (see Figure 18).

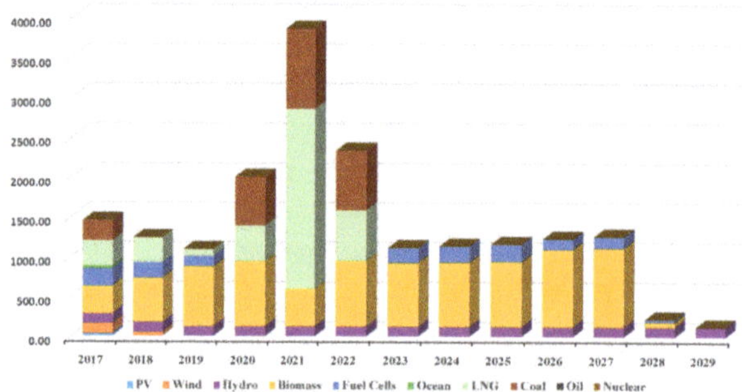

Figure 18. Newly installed capacity of each energy source in SC4.

Next, we checked to see how the RE/NRE generation capacity and power generation changes compare to the reference scenario. In terms of generation capacity and amount of generation, there is no difference from the reference scenario by 2022, but the proportion of RE generation exceeds 10% by 2029. In particular, the proportion of RE capacities decreases but the proportion of RE generation amount increases (see Table 9). This implies that power producers follow different utilization rates for RE/NRE capacity under the same energy mix after reaching grid parity. In this case, the increase in the RE capacity utilization rate is the force that drives up SMP prices. In other words, if grid parity is attained and the RE capacity utilization rate is increased, the profitability deterioration problem of power producers caused by SMPs being too low can be solved. However, when grid parity is attained without adopting the nuclear and coal phase-out option, it is difficult to attain an RE capacity and generation proportion of 20% or more nationally. This implication should be considered when the government intends to increase the RE proportion to achieve sustainability in the national energy mix.

Next, Figure 19 shows how the profit of each segment's representative power producers changed in the same year versus scenario 1. Profits are calculated as profit per MWh to compare on an equal basis, not on a total profit basis. Our results show that the AP segment shows the greatest profit increase per MWh during the planning horizon, as it further strengthened its aggressive investment propensity previously shown in the reference scenario. The IL segment shows a different behavior: a relatively small increase in profits, due to its role in supporting the industry.

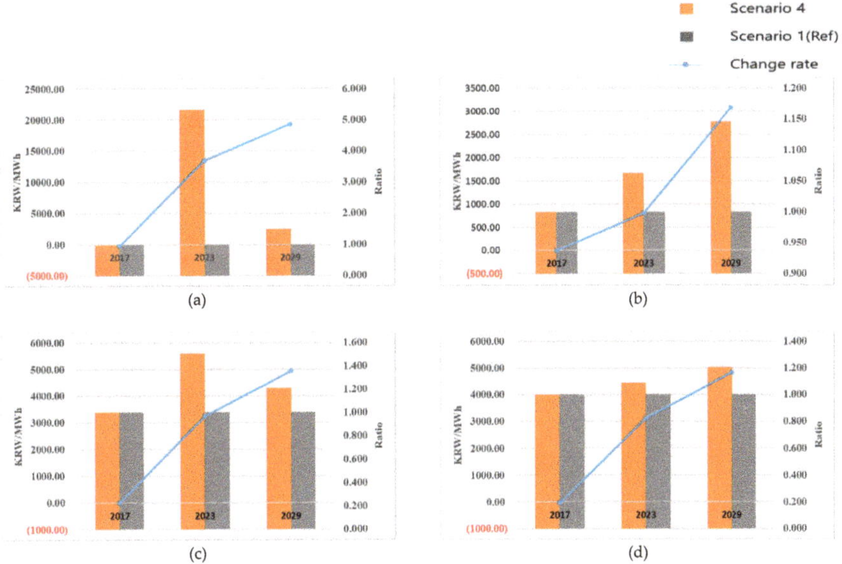

Figure 19. Profit change amount (KRW/MWh) and rate (%) of: (**a**) AP segment (M10); (**b**) IL segment (M1); (**c**) IF segment (M18); (**d**) RS segment (M5) (compared to SC1).

Table 9. Comparison of the ratio of RE/NRE by scenario (SC1 and SC4).

| | Scenario 1 (Ref) | | | | Scenario 4 | | | |
| | % of Electricity Generation Capacity(A) | | % of Electricity Generation(B) | | % of Electricity Generation Capacity(A) | | % of Electricity Generation(B) | |
Year	RE	NRE	RE	NRE	RE	NRE	RE	NRE
2016	2.8	97.2	1.7	98.3	2.8	97.2	1.7	98.3
2022	7.2	92.8	5.9	94.1	7.1	92.9	5.9	94.1
2029	12.6	87.4	9.7	90.3	11.4	88.6	10.2	89.8

6. Discussion and Conclusions

6.1. Discussion and Political Implications

To enable the future power generation industry in Korea to operate in a sustainable and eco-friendly manner, this study derived optimal operational strategies for each power producer under the governmental policies (RPS and ETS). It considered 10 energy sources (six types of RE, four types of NRE) and 18 power producers under both regulations, and developed a mathematical model that included actual policy constraints along with other generation-related constraints. This study formulated the mathematical model intended to maximize the total profit for power producers, and to determine the capacity expansion plans, sales/purchase amounts of REC/CER, and penalty levels. Additionally, a scenario-based analysis was conducted to derive the optimal operational strategy for each power producer based on the scenario design that considered four different influential factors. Furthermore, this study proposed a matrix to analyze the behavior of each power producer in detail. The proposed matrix classified the types of power producers and analyzed how the influential factors of each scenario affect the strategies of power producers in each segment. Based on the analysis, this study drew some meaningful insights. First, the scenario-based analyses show that small hydropower generation adopts steadily, so it is necessary to consider the support policies specializing in small hydropower generation or policies that encourage technological research and the development for expanding realizable potential generation capacity. In addition, in order for the government to more rapidly increase the proportion of RE sources, it is worth targeting the IF segment, to which most power producers belong. Only for those IF segment's power producers that continue to make more favorable decisions between paying penalties and additional capacity investments to accomplish their RPS duties, the government will be able to set policies that increase the penalty rate from the beginning of the regulation to drive IF's decisions toward additional capacity investments. In addition, in the case of the RS and AP segment power producers that are faithfully responding to government regulations, it is possible to consider policies such as loan interest rate benefits and preferential treatment of REC weights depending on the number of successive years of achieving duty.

Plus, in terms of strengthening the carbon emissions regulations, the CER price, which is currently 20,000 KRW per unit CER, does not fulfill the effective regulatory function for reducing the carbon emissions of the power producers. However, the CER price cannot be intentionally increased. Thus, in order to achieve the national goal, the government must be able to establish policies for grading power producers according to the amount of power generation and applying differentiated penalties rate to each grade. Finally, this study expected that attaining grid parity can solve the power producer profitability deterioration problem caused by SMPs that are too low. However, reaching grid parity was not sufficient to achieve RE generation amounts greater than 20%. Therefore, the solution for this problem should contain the nuclear and coal phase-out roadmap, or the development of facility designs and generation technologies that are not biased toward one or two generation sources. In addition, if a power producer invests in energy sources with high investment costs and REC weights, greater support policy should provide for maximizing the effect of reaching grid parity. These insights are important implications that must be considered in order to maximize the effectiveness of the policy-making process and to enact policy that ensures the future sustainability of the national power generation industry. Additionally, in our analysis, one significant conclusion is that, for accomplishing the regulation goals at the national level and for achieving the sustainability of Korea's future power industry, implementing different policies that consider characteristics of each segment is more effective than implementing identical policies for every segment.

6.2. Limitations and Future Research

Despite the above implications, this study has some limitations. First, the research that determines the future energy mix of a country or a company needs to consider the electricity supply and demand uncertainty that changes over time. Our study does not take this into consideration because it mainly aims at analyzing the operational strategies and behaviors of power producers under the RPS and ETS regulations. Second, some assumptions and parameter estimation methods described in this study make the model impractical. In particular, some parameters were inevitably estimated and used in the model due to the limitations regarding available data and data collection.

For future research, it is necessary to overcome these limitations. Such a study must consider uncertainties such as generation and operation costs and price data (for SMP, REC, and CER) to make the model more realistic. In addition, if the technical constraints associated with electric power generation were to further enhance and parameters were to be standardized to make it possible for their more accurate use, the results of our operational strategy optimization model will then become more reliable and more useful. In addition, in order to address various situations from the power producer perspective, it is necessary for the scenario-based analyses to consider fluctuations in electric power demand and fuel prices. Finally, the model could enhance by adding more sophisticated constraints regarding the RPS and ETS regulations.

Author Contributions: D.S. wrote the manuscript and participated in all phases. J.K. participated in scenario-based analysis and reviewed the related literature. B.J. conceived the theme of this study, guided the whole research process, and responded to the reviewers. All authors have read and approved the final manuscript.

Funding: This research received no external funding.

Abbreviations

RPS	Renewable Portfolio Standard
ETS	Emissions Trading Scheme
GHG	Greenhouse Gas
NRE	Non-renewable Energy
RE	Renewable Energy
CER	Carbon Emissions Reductions
REC	Renewable Energy Certificates
GEP	Generation Expansion Planning
BAU	Business as Usual
KEPCO	Korea Electric Power Corporation
EPSIS	Electric Power Statistics Information System
SMP	System Marginal Prices
LNG	Liquefied Natural Gas
IL	Industry Leaders
RS	Reliable Supporters
IF	Industry Followers
AP	Aggressive Participants

Indices

e	Power generation source index ($e = i \cup j$, $e = 1, 2, \dots , 10$)
i	Renewable energy (RE) source index ($I = 1,2, \dots , 6$)
j	Non-renewable energy (NRE) source index ($j = 7, 8, 9, 10$)
m	Power Producer (PP) index ($m = M1, M2, \dots , M18$)
$t(\leftrightarrow k)$	Time Horizon (2017($t = 1$)~2029($t = T = 13$), 2029: Target year)

Decision Variables

$X_{e,m,t}$	New generation capacity of energy source e of each power producer in period t (MW)
$x_{e,m,t,k}$	Constructed new generation capacity of energy source e of each power producer in period k to complete in period t (MW)
$Y_{e,m,t}$	1 if the capacity expansion of energy source e is considered for each power producer in period k to complete in period t ($\in\{0,1\}$)
$QR_{m,t}^{+}$	Sales amount of Renewable Energy Certificate (REC) of each power producer in period t

$QRPU_{m,t}^{-}$	Additional purchasing amount of REC of each power producer in period t
$QRPE_{m,t}^{-}$	Non-fulfillment amount of REC of each power producer in period t
$QC_{m,t}^{+}$	Sales amount of Carbon Emission Reduction (CER) of each power producer in period t
$QCPU_{m,t}^{-}$	Additional purchasing amount of CER of each power producer in period t
$QCPE_{m,t}^{-}$	Non-fulfillment amount of CER of each power producer in period t

Parameters

$add_{e,m,t}$	Planned expansion capacity of energy source e of each power producer in period t (MW)
$Budget_{m,t}$	Budget limitation of each power producer for capacity expansion in period t (KRW)
$Capa_{i\in e,t}^{P}$	Potential expansion capacity of energy source i in period t (MW)
$Capa_{e,m,0}$	Initial generation capacity of power producer of each energy source e (MW)
CF_e	CO_2 emission coefficient of energy source e ($kgCO_2$/MWh)
CUL_m	Upper limit of additional purchasing amount of CER of each power producer in period t (tCO_2)
D_t	Electric power demand in period t(MWh)
Eff_e	Power generation efficiency of energy source e (%)
$ETS_{m,t}^{Quota}$	CO_2 emission limitation of each power producer in period t (tCO_2)
$exp_{e,m,t}$	Planned expired capacity of energy source e of each power producer in period t (MW)
$genc_{e,t}$	Power generation and operation cost of energy source e in period t (KRW/MWh)
Hr_e	Annual generating hours of energy source e (hour)
IC_e	Investment cost of unit expansion capacity of energy source e (KRW/MW)
M	Big M
PC_t	Predicted CER in period t (KRW/MWh)
PE_t	Predicted SMP in period t (KRW/MWh)
PR_t	Predicted REC in period t (KRW/MWh)
RM_t	Reserve margin in period t (%)
$RPS_{m,t}^{Target}$	RPS target ratio of each power producer in period t (%)
RUL_m	Upper limit of additional purchasing amount of REC of each power producer in period t (tCO_2)
$WR_{i\in e}$	Weight of REC issued of renewable energy source i
Γ	Interest rate (%)
δ_e	Available constructed capacity of energy source e in each period (MW/year)
ε	Net ratio of transmission and distribution
Θ	Capacity proportion of power producer (obligatory) compared to total capacity (%)

References

1. Chen, Q.; Kang, C.; Xia, Q.; Zhong, J. Power generation expansion planning model towards low-carbon economy and its application in China. *IEEE Trans. Power Syst.* **2010**, *25*, 1117–1125. [CrossRef]
2. Mitigation of Climate Change. IPCC special report on renewable energy sources and climate change mitigation. *Renew. Energy* **2011**, *20*, 1–24.
3. Kim, S.; Koo, J.; Lee, C.J.; Yoon, E.S. Optimization of Korean energy planning for sustainability considering uncertainties in learning rates and external factors. *Energy* **2012**, *44*, 126–134. [CrossRef]
4. Park, S.Y.; Yun, B.Y.; Yun, C.Y.; Lee, D.H.; Choi, D.G. An analysis of the optimum renewable energy portfolio using the bottom-up model: Focusing on the electricity generation sector in South Korea. *Renew. Sustain. Energy Rev.* **2016**, *53*, 319–329. [CrossRef]
5. Ministry of Trade, Industry, and Energy. *The 2nd National Energy Basic Plan*; Ministry of Trade, Industry, and Energy: Seoul, Korea, 2014.
6. Korea Energy Agency. *Energy Statistics Handbook*; Korea Energy Management Corporation: Ulsan, Korea, 2015.
7. Ministry of Environment. *National Emission Allocation Plan*; Ministry of Environment: Seoul, Korea, 2014.
8. Dehghan, S.; Amjady, N.; Kazemi, A. Two-stage robust generation expansion planning: A mixed integer linear programming model. *IEEE Trans. Power Syst.* **2014**, *29*, 584–597. [CrossRef]
9. Li, S.; Coit, D.W.; Felder, F. Stochastic optimization for electric power generation expansion planning with discrete climate change scenarios. *Electr. Power Syst. Res.* **2016**, *140*, 401–412. [CrossRef]

10. Valinejad, J.; Barforoushi, T. Generation expansion planning in electricity markets: A novel framework based on dynamic stochastic MPEC. *Int. J. Electr. Power Energy Syst.* **2015**, *70*, 108–117. [CrossRef]
11. Bhattacharya, A.; Kojima, S. Power sector investment risk and renewable energy: A Japanese case study using portfolio risk optimization method. *Energy Policy* **2012**, *40*, 69–80. [CrossRef]
12. Schelly, C. Implementing renewable energy portfolio standards: The good, the bad, and the ugly in a two state comparison. *Energy Policy* **2014**, *67*, 543–551. [CrossRef]
13. Upton, G.B.; Snyder, B.F. Renewable energy potential and adoption of renewable portfolio standards. *Util. Policy* **2015**, *36*, 67–70. [CrossRef]
14. Barbose, G.; Wiser, R.; Heeter, J.; Mai, T.; Bird, L.; Bolinger, M.; Carpenter, A.; Heath, G.; Keyser, D.; Macknick, J.; et al. A retrospective analysis of benefits and impacts of US renewable portfolio standards. *Energy Policy* **2016**, *96*, 645–660. [CrossRef]
15. Bowen, W.M.; Park, S.; Elvery, J.A. Empirical estimates of the influence of renewable energy portfolio standards on the green economies of states. *Econ. Dev. Quart.* **2013**, *27*, 338–351. [CrossRef]
16. Yoon, J.H.; Sim, K.H. Why is South Korea's renewable energy policy failing? A Qual. Eval. *Energy Policy* **2015**, *86*, 369–379. [CrossRef]
17. Kwon, T.H. Is the renewable portfolio standard an effective energy policy?: Early evidence from South Korea. *Util. Policy* **2015**, *36*, 46–51. [CrossRef]
18. Choi, D.G.; Park, S.Y.; Hong, J.C. Quantitatively exploring the future of renewable portfolio standard in the Korean electricity sector via a bottom-up energy model. *Renew. Sustain. Energy Rev.* **2015**, *50*, 793–803. [CrossRef]
19. Calvo-Silvosa, A.; Antelo, S.I.; Soares, I. The European low-carbon mix for 2030: The role of renewable energy sources in an environmentally and socially efficient approach. *Renew. Sustain. Energy Rev.* **2015**, *48*, 49–61.
20. Yi, B.W.; Xu, J.H.; Fan, Y. Coordination of policy goals between renewable portfolio standards and carbon caps: A quantitative assessment in China. *Appl. Energy* **2019**, *237*, 25–35. [CrossRef]
21. Jiang, J.J.; Ye, B.; Ma, X.M. The construction of Shenzhen's carbon emission trading scheme. *Energy Policy* **2014**, *75*, 17–21. [CrossRef]
22. Wu, L.; Qian, H.; Li, J. Advancing the experiment to reality: Perspectives on Shanghai pilot carbon emissions trading scheme. *Energy Policy* **2014**, *75*, 22–30. [CrossRef]
23. Park, H.; Hong, W.K. Korea's emission trading scheme and policy design issues to achieve market-efficiency and abatement targets. *Energy Policy* **2014**, *75*, 73–83. [CrossRef]
24. Zhang, X.; Qi, T.Y.; Ou, X.M.; Zhang, X.L. The role of multi-region integrated emissions trading scheme: A computable general equilibrium analysis. *Appl. Energy* **2017**, *185*, 1860–1868. [CrossRef]
25. Hong, Z.; Chu, C.; Zhang, L.L.; Yu, Y. Optimizing an emission trading scheme for local governments: A Stackelberg game model and hybrid algorithm. *Int. J. Prod. Econ.* **2017**, *193*, 172–182. [CrossRef]
26. Boersen, A.; Scholtens, B. The relationship between European electricity markets and emission allowance futures prices in phase II of the EU (European Union) emission trading scheme. *Energy* **2014**, *74*, 585–594. [CrossRef]
27. Song, T.H.; Lim, K.M.; Yoo, S.H. Estimating the public's value of implementing the CO_2 emissions trading scheme in Korea. *Energy Policy* **2015**, *83*, 82–86. [CrossRef]
28. Li, F.; Schwarz, L.; Haasis, H.D. A framework and risk analysis for supply chain emission trading. *Logist. Res.* **2016**, *9*, 10. [CrossRef]
29. Shim, S.; Lee, J. On Inefficiency Factors in the Korean Emissions Trading Scheme. *Korea Energy Econ. Rev.* **2015**, *14*, 177–211.
30. Hemmati, R.; Hooshmand, R.A.; Khodaabakhshian, A. Reliability constrained generation expansion planning with consideration of wind farms uncertainties in deregulated electricity market. *Energy Convers. Manag.* **2013**, *76*, 517–526. [CrossRef]
31. Koltsaklis, N.E.; Georgiadis, M.C. A multi-period, multi-regional generation expansion planning model incorporating unit commitment constraints. *Appl. Energy* **2015**, *158*, 310–331. [CrossRef]
32. Antunes, C.H.; Martins, A.G.; Brito, I.S. A multiple objective mixed integer linear programming model for power generation expansion planning. *Energy* **2004**, *29*, 613–627. [CrossRef]
33. Gitizadeh, M.; Kaji, M.; Aghaei, J. Risk based multi-objective generation expansion planning considering renewable energy sources. *Energy* **2013**, *50*, 74–82. [CrossRef]

34. Muñoz-Delgado, G.; Contreras, J.; Arroyo, J.M. Joint expansion planning of distributed generation and distribution networks. *IEEE Trans. Power Syst.* **2015**, *30*, 2579–2590. [CrossRef]
35. Luz, T.; Moura, P.; de Almeida, A. Multi-objective power generation expansion planning with high penetration of renewables. *Renew. Sustain. Energy Rev.* **2018**, *81*, 2637–2643. [CrossRef]
36. Ghaderi, A.; Moghaddam, M.P.; Sheikh-El-Eslami, M.K. Energy efficiency resource modeling in generation expansion planning. *Energy* **2015**, *68*, 529–537. [CrossRef]
37. Li, S.; Tirupati, D. Dynamic capacity expansion problem with multiple products: Technology selection and timing of capacity additions. *Oper. Res.* **1994**, *42*, 958–976. [CrossRef]
38. Chen, X.; Lv, J.; McElroy, M.B.; Han, X.; Nielsen, C.P.; Wen, J. Power system capacity expansion under higher penetration of renewables considering flexibility constraints and low carbon policies. *IEEE Trans. Power Syst.* **2018**, *33*, 6240–6253. [CrossRef]
39. Lim, H.; Jo, H. Analysis on the effects of RPS and FIT policies on the renewable energy supply: Panel tobit analysis of 104 countries. *Korean Energy Econ. Rev.* **2017**, *16*, 1–31.
40. Lee, K.; Kim, L.; Yeo, Y. Optimization of Integrated District Heating System (IDHS) Based on the Forecasting Model for System Marginal Prices (SMP). *Korean Chem. Eng. Res.* **2012**, *50*, 479–491. [CrossRef]
41. Ministry of Knowledge Economy. *The 7th Basic Plan of Long-Term Electricity Supply and Demand*; Ministry of Knowledge Economy: Seoul, Korea, 2015.
42. Ahn, J.; Woo, J.; Lee, J. Optimal allocation of energy sources for sustainable development in South Korea: Focus on the electric power generation industry. *Energy Policy* **2015**, *78*, 78–90. [CrossRef]
43. Korea Power Exchange. Electric Power Statistics Information System (EPSIS). Available online: http://epsis.kpx.or.kr/ (accessed on 15 November 2017).
44. Moné, C.; Smith, A.; Maples, B.; Hand, M. *2013 Cost of Wind Energy Review*; NREL Technical Report TP-5000-63267; National Renewable Energy Laboratory: Lakewood, CO, USA, 2015.
45. International Energy Agency. *Projected Costs of Generating Electricity*, 2015th ed.; International Energy Agency: Paris, France, 2015.
46. Lazard. *Levelized Cost of Energy Analysis–Version 10.0*; Lazard: Hamilton, Bermuda, 2016.
47. New and Renewable Energy Data Center. National Potential Statistics. Available online: http://kredc.kier.re.kr/kier/03_dataEnquiry/Whole_measure_static.aspx (accessed on 10 November 2017).
48. Korea Exchange. CER Market Price Information. Available online: http://marketdata.krx.co.kr/mdi#document=070301 (accessed on 6 November 2017).
49. Tidball, R.; Bluestein, J.; Rodriguez, N.; Knoke, S. *Cost and Performance Assumptions for Modeling Electricity Generation Technologies*; National Renewable Energy Lab: Golden, CO, USA, 2010.

Article

The International Energy Security Risk Index in Sustainable Energy and Economy Transition Decision Making—A Reliability Analysis

Iztok Podbregar [1], Goran Šimić [2], Mirjana Radovanović [3], Sanja Filipović [4], Damjan Maletič [5] and Polona Šprajc [1,*]

1   Department of Organization and Management, Faculty of Organizational Sciences, University of Maribor, Kidričeva cesta 55a, Kranj 4000 Slovenia; iztok.podbregar@um.si
2   Department for Simulations and Distance Learning, University of Defense in Belgrade, Pavla Jurišića Šturma 1, Belgrade 11000, Serbia; gshimic@gmail.com
3   Department of Energy Security, Faculty of Security Studies, Educons University, 87 Vojvode Putnika Street, Sremska Kamenica 21208, Serbia; mirjana4444@gmail.com
4   Department of Economic Theory and Analysis, Faculty of Business, Singidunum University, Danijelova 32, Belgrade 11000, Serbia; sfilipovic@singidunum.ac.rs
5   Department of Enterprise Engineering, Faculty of Organizational Sciences, University of Maribor, Kidričeva cesta 55a, Kranj 4000, Slovenia; damjan.maletic@um.si
*   Correspondence: polona.sprajc@um.si; Tel.: +386-40-802-222

Received: 28 June 2020; Accepted: 16 July 2020; Published: 17 July 2020

Abstract: The world economy and society are in a complex process of transition characterized by a high degree of uncertainty. Therefore, further development and management of the transition will largely depend on the quality of the decisions made and, accordingly, on the decision-making process itself. The main goal of this study is to analyze the reliability of International Energy Security Risk Index as a tool to support the process of energy and economy transition decision making, as closely related and highly interdependent phenomena. The index is composed of 29 aggregated variables (grouped into eight categories), and the research is conducted on a research sample of 25 countries over a period of 36 years. The reliability assessment is performed by using Multiple Regression Analysis. Multicollinearity test, plus Multicollinearity test with Variance Inflation Factors, is used for methodological verification. The test results indicate a high degree of unreliability of the Index, as is concluded based on the observed errors in its methodological settings. These errors primarily relate to a high degree of multicollinearity in all 29 variables, whereby independent variables lose their independence and thus jeopardize reliability of the total Index. Out of the eight groups of variables, the fuel imports group is the only one that does not show big methodological errors. The paper presents a recommendation for the improvement of the observed Index (review of the role of individual variables found to be particularly methodologically indicative), as well as a recommendation for different distribution of weighting coefficients.

Keywords: energy and economy transition; decision making; international energy security risk index; multiple regression analysis; multicollinearity test; multicollinearity test with variance inflation factors

---

1. Introduction

Energy, geopolitical and economic transition, along with changes in other fields of development, represents a great challenge facing humanity in the XXIst century [1]. In such a situation, decision making, as an already complex and demanding process, becomes even more difficult because it is associated with a high degree of uncertainty on several grounds [2]. Firstly, decisions must be made in the unforeseen circumstances that characterize the present time, with extremely difficult possibilities for predicting future events: geopolitical turmoil at the global level (as well as in some countries and regions), economic crises and problems at the micro and macro levels, the consequences caused by the Covid-19 virus pandemic, social tensions and unrest, changes in the field of security, resource exploitation, environmental concerns, climate change challenges and the like. Decision making in conditions of such a high degree of uncertainty, risk and ambiguity, is an extremely demanding and very important issue that calls for adequate solutions by taking into account specific circumstances [3]. Therefore, these solutions cannot be universal [4]. Secondly, decision making is hampered by the fact that decisions can be made in different ways, while using different approaches: from extremely subjective to milti-attribute decision making, which is supported by the objective fact that some decisions must be made in an extremely short time, caused by certain emergencies. Therefore, robust decision-making approaches are useful, yet, studies on them are rare [5]. In addition, there is no consensus on or uniform guidance about which decision-making method should be used under what conditions or for what purpose [6].

Since energy and economy transition takes place in the same conditions, with positive causal relationship [7]. Energy and macroeconomic variables are highly connected and evaluated in scientific literature [8]. Upon analysing the relation between energy security and macroeconomic variables in European countries, it was found that increase of Gross Domestic Product (GDP) is positively correlated, while the Consumer Price Index (CPI) is negatively correlated with energy security [9]. Analysing the relation between energy security and five macroeconomic variables (GDP, inflation, current account balance, foreign direct investment, and employment) in EU countries, Granger causality tests of the panel VAR model reveal that in the short run only employment may be negatively affected by energy security. Thess results may suggest to EU policy makers that the focus of energy policy should be primarly devoted to technological advances and energy consumption efficiency [10]. Empirical research based on 10 Asian economies provides evidence that energy price volatility has a positive impact on GDP in the short-run and the opposite in the long-run [11]. Another study [12] confirmed that increases in oil prices have a lower effect on GDP growth the lower the oil-to-energy ratio. These results are in line with the hypothesis that a higher energy import dependency makes a country more vulnerable to price variations. Facing a decline of oil-to-energy ratios, the decreasing influence of oil price fluctuations on GDP growth during the past 40 years can be explained. Emprirical research in which the impact of unexpected oil price changes on macroeconomic indicators in 19 major oil-related countries was analysed [13] showed that the macroeconomic overview of energyiimporting countries is under the higher pressure due to the fact they must pay more attention to the oil price shock response system. Therefore, further research should be implemented not only to analyse the relation between energy and macroeconomic indices, but also to design advanced indicators of energy security [14,15].

Taking into consideration the strong connection between energy and economy transition, decision making in this field is also complex and has certain additional specifics [16]. Namely, every country strives to adapt to the conditions of transition and, at the same time, to achieve and maintain a satisfactory level of economic stability and energy security for as long as possible, which allows it to create important preconditions for economic development, political stability and quality of life. In addition, transition decision making should take into account a large number of parameters [17], with specifics to be considered when it comes to developed or developing countries [18], among which many have been changed and aligned with the requirements of sustainable development [19]. Decisions in this area are more complex when developing a common energy policy in countries that have their own differences [20].

The decision-making process in the energy security area is further aggravated for several reasons. First of all, energy security is not defined in a unique way and, therefore, methodologies for measuring energy security can be considered questionable. However, with the increasing importance of energy security, efforts to solve this problem to some extent are growing as well. Therefore, a number of approaches and indices for measuring energy security have been developed and measurements have been carried out in individual countries, regions or globally, with the aim of determining the situation, defining the trends and predicting the future [21].

Energy security can be viewed in different ways, but for the purpose of decision making it is necessary to first clearly define the parameters that make up energy security [22]. On this point, there are already different views and analyses made on the basis of energy security concepts [23] that are set differently, i.e., insufficiently defined relations between the dimensions that make up energy security [24], with differences in conceptualization for decision with long and short-term impact [25]. The European Commission agrees as it defines energy security as the "uninterrupted physical availability of energy products on the market, at a price which is affordable for all consumers (private and industrial), while respecting environmental concerns and looking towards sustainable development" [26], whereby it considers availability and accessibility of supply, economic affordability and environmental sustainability to be basic dimensions.

Energy security can be viewed as an indicator of availability, accessibility, affordability, and acceptability [27], whereby it is proposed to add fifth dimension (efficiency) to the four dimensions mentioned [28]. In the Energy Trilemma Index case (World Energy Council), energy security is observed through the unity of four dimensions: energy security, environmental sustainability, energy equity and country context [29]. Efforts are also evident to develop special measurement methods for countries with certain specificities. Thus, in the case of energy import dependent countries that are, at the same time, large consumers, energy security is viewed as a unity of vulnerability, efficiency and sustainability [30]. In the case of highly industrialized countries, it is recommended to use a four-dimensional energy security model: availability, affordability, energy efficiency, and environmental stewardship [31]. Different combination of indicators is used in case of assessing energy security in Baltic States. In this case, combination of technical, economic, geopolitical and sociopolitical aspects of energy security has been used [32]. On the other hand, energy security should be planned and implemented by respecting sustainability prerequisites [33]. Energy security could be evaluated as the ability of the energy system to resist disruptions [34]. It can be concluded that the list of dimensions that make up energy security is continuously expanding. At the same time, there is an emphasized impossibility of developing all dimensions in parallel and simultaneously, and it further complicates decision making [35].

By analyzing the literature, it can be said with a high degree of certainty that a single definition will probably never be adopted. However, it is much more important to analyze the existing methods and to find ways to improve and adapt them [36].

After analyzing 63 studies on measuring energy security, some conclusions have been defined and they can, in a sense, serve as directions of further research. For the purposes of this paper, the selected conclusions relate to the assessment of the reliability of the methods applied. A multivariate analysis (usually in the form of Principal Component Analysis) was performed in only 10 studies, while more complex sensitivity and robustness analyses are even rarer. "During the process of building an index, many decisions are made, such as the selection of the indicator set, data normalization, indicator weighting and aggregation. The validity of these decisions and the resulting indices may thus be contested. Uncertainty, sensitivity and robustness analysis support the confidence into an index and its underlying policy messages". Regardless of its importance, this analysis was performed in only three studies. In summary, out of a total of 63 studies, certain assessment of the methodological settings of the index under observation, and analysis of its reliability, was conducted in only 13 cases [37].

Due to all of the above, and primarily due to the great importance of decision making in energy transition, with an emphasis on energy security as an important goal of energy and economic transition strategies, a study on the International Energy Security Risk Index (hereinafter IESRI), created by the U.S. Chamber of Commerce, is conceived and implemented in this paper, with an emphasis on the analysis of its methodological settings, the variables used, the assessment of correlation between them, the assessment of multicollinearity—all with the aim of drawing a conclusion on its reliability for the application in energy transition decision making, and with recommendations for improvement.

The paper is divided in five sections. The Introduction section presents the background, subject and goal of the research, as well as a review of the literature related to it. In the Material and Methods section, the research is presented, the sample is defined and the methods used are described. The results of the research are presented in the Results section. The interpretation of the obtained results, along with certain comments and recommendations, are provided in the Discussion section. Finally, an overview of the study, its basic results and interpretations are provided in the Conclusions section, along with indications of the directions of further research.

2. Materials and Methods

A reliability assessment in this research was conducted for International Index of Energy Security Risk which tracks variables in the field of energy security in 25 selected countries, grouped per regions [38]:

The European region includes 11 countries: Denmark, Germany, France, Italy, Norway, The Netherlands, Russian Federation, Poland, Spain, Ukraine and United Kingdom.

The Americas includes four countries: Brazil, Mexico, Canada and United States of America.

The Asian region is represented by China, India, Indonesia, Japan, South Korea, Thailand and Turkey.

Australia and New Zealand.

The only country from Africa represented in this study is South Africa.

The International Index of Energy Security Risk consists of eight groups of variables, presented as follows:

- The *Global fuels* group is represented by six variables: Global oil reserves, Global gas reserves, Global coal reserves, Global oil production, Global gas production and Global coal production.
- *Fuel imports* consists of five variables: Oil import exposure, Coal import exposure, Gas import exposure, Fossil fuel import expenditure per GDP and Total energy import exposure.
- *Energy expenditures* are explained with four variables: Crude oil prices, Retail electricity prices, Energy expenditure intensity and Energy expenditure per capita.
- *Price and market volatility*: GDP per capita, Energy expenditure volatility, Crude oil price volatility and World oil refinery usage.
- *Energy use intensity* includes Energy intensity, Energy consumption per capita and Petroleum intensity.
- *Electric power sector* is described with two variables: Electricity diversity and Non-carbon generation.
- *Transportation sector*, includes two variables: Transport energy intensity and Transport energy per capita.
- *Environmental* group is presented with three variables: CO_2 emission trend, CO_2 GDP intensity and CO_2 per capita.

The research was conducted for the period from 1980 to 2016 with data updated for that period, and the methods used in this paper are:

- Multiple Regression Analysis, selected because it allows one to define the contribution of each selected predictor to the level of the variance of the total Index [39], which is the first important assumption for the analysis of the methodological settings of the observed index.
- The Multicollinearity test is used to determine whether and to what extent the independent variables under observation are correlated. The occurrence of a high degree of correlation is certainly a significant methodological problem, because independent variables thus lose their presumption of independence as one of the most important preconditions for their use.
- A Multicollinearity test with Variance Inflation Factors (VIF) was conducted in addition to a Multicollinearity test, with the aim of establishing whether there is a real possibility to reduce

multicollinearity to a satisfactory framework. The VIF value between 1 and 5 indicates a correlation, but not significant enough to be subject to correction. If the VIF value is greater than 5, it indicates a high degree of multicollinearity and shows that methodological corrections are necessary [40].

The research started from the hypothesis that the existing methodology for IESRI assessment is too complex and that it can be simplified. In each of the six groups special metrics were developed to express the values of its variables. Not all of the variables from one group have the same importance. This difference is expressed by weighting factors (weights) assigned to them. Thus, the importance of a group represents the cumulative value of weighting factors of its parameters.

In order to examine the impact of the parameters on IESR, Multiple Regression Analysis was performed first. Given a large number of parameters which together describe 100% of IESRI, the verification of methodological correctness was performed by using a Multicollinearity test. The main purpose of this test is to detect mutual corelations among the IESRI parameters as independent variables. The large number of variables used in model indicated the necessity of the Multicollinearity test. Variance Inflation Factor (VIF) and Tolerance were used as measures to detect multicollinearity. We used both of them as they have reciprocal nature based on coefficient of determination (R). Both factors express a degree of contribution of observed parameter to the standard error. Emphirically, if the VIF values are less than 10, and the Tolerance values are above 0.1, cross-correlation of variables is considered to be within the tolerance, i.e., variables are considered as independent. In other words, if the VIF values are 10 and higher and the Tolerance values are less than 0.1, in that case there is a large cross-correlation between the variables, i.e., not all variables are independent.

3. Results

The formation of IESRI from variables of unequal importance, as determined by affiliation to a group and by weighting factors assigned, may seem like a satisfactory solution, however, when all variables are observed together (since only in this way do they determine the index completely to 100%), multicollinearity or mutual overlap (correlation) between the variables is much higher than allowed. This finding was confirmed by the Multicollinearity test.

Table 1 provides a set of the variables, their importance for the formation of IESRI, and their cross-correlation. Presented coefficients (Beta, T-test, Sig, Partial, VIF) are important for reliable determining candidates (parameters) for removing from the model (shown in Bold). The Beta coefficient determines how changes of predictor parameters influence on change of IESRI as a dependent variable. It is important only if the T-test value is greater than 2. Then then Beta coefficient has satisfactory significance for further consideration about the predictor. A significance or p-value (*Sig*) greater than 0.05 shows that there is no significant partial correlation with IESRI (dependent variable) and observed parametar is a candidate for exclusion.

Table 1. Multicollinearity and partial correlation without influence on IESRI.

Parameter	Beta	T test	Sig	Partial	VIF
(Constant)		−0.702	0.497		
Global_Oil_Reserves	0.044	8.685	0.000	0.934	32.369
Global_Oil_Production	**0.022**	**1.078**	**0.304**	**0.309**	**527.363**
Global_Gas_Reserves	0.037	4.185	0.002	0.784	96.703
Global_Gas_Production	0.102	5.880	0.000	0.871	376.293
Global_Coal_Reserves	0.043	2.752	0.019	0.639	296.586
Global_Coal_Production	0.060	3.154	0.009	0.689	448.817
Oil_Import_Exposure	0.037	2.617	0.024	0.619	252.046
Gas_Import_Exposure	0.136	13.090	0.000	0.969	133.207
Coal_Import_Exposure	0.160	10.640	0.000	0.955	280.947

Table 1. *Cont.*

Parameter	Beta	T test	Sig	Partial	VIF
Transport_Energy_Intensity Exposure	**0.037**	**1.745**	**0.109**	**0.466**	**558.111**
Energy_Expenditure_Intensity	0.224	4.045	0.004	0.677	335.98
Fossil_Fuel_Import_Expenditure_per_GDP	0.096	2.870	0.015	0.654	1390.620
Energy_Expenditure per Capita	0.123	3.477	0.005	0.724	1543.139
Crude_Oil_Prices	0.305	29.979	0.000	0.994	128.419
Retail_Electricity_Prices	0.074	10.958	0.000	0.957	57.219
Crude_Oil_Price_Volatility	0.283	79.371	0.000	0.999	15.807
Energy_Expenditure_Volatility	0.144	32.418	0.000	0.995	24.645
World_Oil_Refinery_Utilization	0.025	2.592	0.025	0.616	118.180
GDP per Capita	**0.099**	**1.168**	**0.267**	**0.332**	**8862.667**
Energy_Consumpiton per Capita	0.059	3.993	0.002	0.769	266.549
Petroleum_Intensity	**0.071**	**1.386**	**0.193**	**0.386**	**3234.543**
Energy_Intensity	0.108	2.998	0.009	0.665	198.808
Non-Carbon Generation	**0.002**	**0.337**	**0.743**	**0.101**	**66.905**
Electricity_Diversity	0.062	5.247	0.000	0.845	172.240
Transport_Energy_Intensity	0.102	2.287	0.001	0.545	76.555
Transport_Energy per Capita	0.104	3.037	0.011	0.675	1462.486
CO_2 per Capita	**−0.043**	**−0.657**	**0.525**	**−0.194**	**5299.321**
CO_2_Emission_Trend	0.224	1.892	0.667	0.747	112.003
CO_2 GDP Intensity	**0.178**	**2.076**	**0.062**	**0.530**	**9160.721**

The partial correlation indicator with dependent variable (Partial) indicates a very low contribution of all variables to the formation of the index, given that the partial correlation with dependent variable (IESRI average value for OECD countries) is statistically insignificant ($p > 0.05$). In addition, there is a problem of mutual collinearity a large degree of overlap with other independent variables (the VIF indicator significantly higher than 10). Based only on the value of VIF it is noticed that none of the 29 variables used in the calculation of IESRI meet the condition of independence. Variables with high VIF value, significance (Sig) greater than 0.05 (which shows that the partial correlation with IESRI as dependent variable is not significant) and low significance of the Beta coefficient due to low value of T-test are marked and they represent candidates for possible exclusion from the model.

Given an excessively high degree of correlation between independent variables on the one hand, as well as their low contribution to the formation of the total IESRI on the other hand, the procedure was repeated at the level of all 8 groups of variables. The goal of this approach is not to change the model based on the entirety but based on an insight gained into the behavior of individual variables within the group to which they belong. This refers to an insight into whether some variables are redundant and whether they can be replaced by other variables without significant losses after the formation of IESRI. The following sections describe the groups and associated variables in more detail.

3.1. Global Fuels

Based on the regression analysis, six variables from the Global Fuels group jointly show a high coefficient of determination for this group in relation to the linear model being described. Their contribution to the formation of the total IESRI is almost 87.90%, as presented in Table 2.

Table 2. The result of Multiple Regression Analysis for six parameters from the Global Fuels category.

Model	R	R Square	Adjusted R Square	Std. Error Estimate
1	0.938	0.879	0.855	28.82304

Multicollinearity and partial correlation of parameters from this category are shown in Table 3. Constant is included as an artificial parameter (a technical part of the statistical model) that provides zero mean of residuals. It represents the response value in case of zero values of predictor variables.

Constant provides the consistent regression line. Without it, the regression line will not fit the model. In other words, the absence of Constant will produce additional bias of regression coefficients used in analysis.

Table 3. Multicollinearity and partial correlation of the parameters from the Global Fuels group.

Parameter		Beta	T test	Sig	Partial	Tolerance	VIF
(Constant)		−0.579	0.567				
Global_Oil_Reserves	**0.304**	**1.458**	**0.155**	**0.257**	**0.093**	**10.782**	**10.782**
Global_Oil_Production	**−0.162**	**−0.715**	**0.480**	**−0.129**	**0.078**	**12.803**	**12.803**
Global_Gas_Reserves	−0.309	−2.027	0.052	−0.347	0.174	5.763	5.763
Global_Gas_Production	−0.783	−3.291	0.003	−0.515	0.071	14.092	14.092
Global_Coal_Reserves	**2.281**	**4.156**	**0.000**	**0.604**	**0.013**	**74.894**	**74.894**
Global_Coal_Production	**0.916**	**4.454**	**0.000**	**0.631**	**0.095**	**10.526**	**10.526**

In contrast to the summary results presented in Table 1, statistical indicators (primarily the Beta weight) at the level of this group show a greater differentiation of the importance of parameters for the formation of IESRI. For example, the Global Coal Reserves and Global Coal Production parameters are of much greater importance for the formation of IESRI values (because they show the highest Beta values), while the Global Oil Reserves and Global Oil Production parameters (shown nin Bold) are at the level of redundant parameters and are almost insignificant for the formation of IESRI (in addition to smaller Beta values, they also show a high degree of overlap with other parameters from the same category (Tolerance values < 1, VIF > 10)).

Based on the Pearson correlations of parameters from the Global Fuels group, the problem of multicollinearity is noticeable in Global Coal Reserves-GCR, Global Oil Production-GOP, Global Gas Production-GGP, Global Gas Reserves-GGR Table 4, where the values of Pearson coefficient are greater than limit of 0.7 (shown in Bold), as shown in Table 4.

Table 4. Pearson Correlation parameters from the *Global Fuels* group.

	IESRI	GOR	GOP	GGR	GGP	GCR	GCP
IESRI	1.000	−0.739	0.693	0.464	0.470	0.648	0.274
Global_Oil_Reserves	−0.739	1.000	−0.314	−0.289	−0.257	−0.404	−0.505
Global_Oil_Production	0.693	−0.314	1.000	**0.701**	**0.785**	**0.887**	−0.263
Global_Gas_Reserves	0.464	−0.289	**0.701**	1.000	**0.876**	**0.837**	−0.272
Global_Gas_Production	0.470	−0.257	**0.785**	**0.876**	1.000	**0.928**	−0.424
Global_Coal_Reserves	0.648	−0.404	**0.887**	**0.837**	**0.928**	1.000	−0.416
Global_Coal_Production	0.274	−0.505	−0.263	−0.272	−0.424	−0.416	1.000

It can be noted that the greatest correlation exists between the Global_Gas_Production and Global_Coal_Reserves variables (0.928). Cross-correlation of these variables indicates the need to separate these variables, i.e., not to place them in the same group (if the above proves to be justified).

3.2. Fuel Imports

The Fuel Imports category contains the most important parameters, given that their contribution to the formation of the total IESRI is 92.9%. The results of the analysis of this group of variables are presented in Table 5.

Table 5. The result of Multiple Regression Analysis for parameters from the Fuel Imports group.

Model	R	R Square	Adjusted R Square	Std. Error Estimate
1	0.964	**0.929**	0.917	21.78287

In addition to satisfactory R Square values (0.929), no special methodological problems were observed regarding the Fuel Imports group, as was confirmed by multicollinearity analysis and presented in Table 6.

Table 6. Multicollinearity and partial correlation of parameters from the Fuel Imports group.

Parameter	Beta	T test	Sig	Partial	Tolerance	VIF
(Constant)		4.004	0.000			
Total_Energy_Import_Exposure	−0.157	−2.362	0.025	−0.391	0.518	1.929
Gas_Import_Exposure	0.285	3.089	0.004	0.485	0.270	3.699
Oil_Import_Exposure	0.251	2.246	0.032	0.374	0.183	5.458
Coal_Import_Exposure	0.361	3.396	0.002	0.521	0.204	4.910
Fosil_Fuel_Import_Expenditure per GDP	0.545	4.415	0.000	0.621	0.151	6.639

The results of this analysis show that the variables that make up this group do not overlap significantly (Tolerance min = 0.151, VIFmax = 6.639). In accordance with the obtained results, it can be said that there is no need for methodological changes in variables from the *Fuel Imports* group.

3.3. Energy Expenditure

The Energy Expenditure category contains variables that contribute 84.4% to the variance of the total IESRI (84%). The results of Multiple Regression Analysis for this group are shown in Table 7.

Table 7. The result of Multiple Regression Analysis for parameters from the Energy Expenditure group.

Model	R	R Square	Adjusted R Square	Std. Error Estimate
1	0.918	**0.844**	0.824	31.77544

Further analysis of parameters from the Energy Expenditure category and, as verified, multicollinearity test, show their polarized nature as presented in Table 8.

Table 8. Multicollinearity and partial correlation of parameters from the Energy Expenditure group.

Parameter	Beta	T test	Sig	Partial	Tolerance	VIF
(Constant)		9.94	0.000			
Energy_Expenditure_Intensity	**0.371**	**3.524**	**0.001**	**0.529**	**0.440**	**2.270**
Energy_Expenditure per Capita	−0.032	−0.157	0.0877	−0.028	0.115	8.674
Retail_Electricity_Prices	0.150	1.553	0.130	0.265	0.526	1.902
Crude_Oil_Prices	**0.597**	**2.908**	**0.007**	**0.457**	**0.116**	**8.621**

Although there is no methodological problem of multicollinearity, the low value of Beta coefficient in the Energy Expenditures per Capita and Retail Electricity Prices parameters indicates that they have no importance within this group for the formation of IESRI values. Therefore, their contribution to IESRI formation should be further considered in more detail.

3.4. Price and Market Volatility

As shown in Table 9, parameters from the Price and Market Volatility category are important for the formation of the total IESRI (77.5%).

Table 9. The result of Multiple Regression Analysis for parameters from the Price and Market Volatility group.

Model	R	R Square	Adjusted R Square	Std. Error Estimate
1	0.880	**0.775**	0.747	38.12479

The multicollinearity test results for this group of variables are given in Table 10.

Table 10. Multicollinearity and partial correlation of parameters from *the Price and Market Volatility* group.

Parameter	Beta	T test	Sig	Partial	Tolerance	VIF
(Constant)		3.877	0.000			
Crude_Oil_Price_Volatility	**0.476**	**3.426**	**0.002**	**0.518**	**0.364**	**2.748**
Energy_Expenditure_Volatility	**0.310**	**2.025**	**0.051**	**0.337**	**0.301**	**3.322**
World_Oil_Refinery_Utilization	−0.389	−10.960	0.059	−0.327	0.178	5.607
GDP per Capita	−0.166	−0.847	0.403	−0.148	0.184	5.447

It is noticeable that the GDP per Capita and World Oil Rafinery Utilization variables do not contribute significantly to the formation of the total IESRI (the significance of World Oil Rafinery Utilization is at the limit value of 0.05, while it is far below the significance limit for GDP per Capita). None of the variables in this category have a multicollinearity problem. It is evident that the Crude Oil Price Volatility and Energy Expenditure Volatility variables have a significant impact on the formation of IESRI.

3.5. Energy Use Intensity

Multicollinearity test for the Energy Use Intensity group shows that variables from this group contribute the least to the formation of the total IESRI (40.4%, shown in Bold), as shown in Table 11.

Table 11. The result of Multiple Regression Analysis for parameters from the Energy Use Intensity group.

Model	R	R Square	Adjusted R Square	Std. Error Estimate
1	0.636	**0.404**	0.350	61.07523

Further analysis showed that only one of the three parameters (Energy Consumption per Capita) statistically significantly (Sig value, shown in Bold) contributes to the formation of IESRI, while Energy Intensity and Petroleum Intensity do not make a significant contribution, as shown in detail in Table 12.

Table 12. Multicollinearity and partial correlation of parameters from the Energy Use Intensity group.

Parameter	Beta	T test	Sig	Partial	Tolerance	VIF
(Constant)		2.646	0.012			
Energy_Consumption per Capita	−0.563	−3.121	**0.004**	−0.477	0.554	1.805
Energy_Intensity	2.710	1.701	0.098	0.284	**0.007**	**140.643**
Petroleum_Intensity	−2.969	−1.916	0.064	−0.316	**0.008**	**132.928**

As regards the Energy Intensity and Petroleum Intensity parameters, a significant problem is evident in multicollinearity which is reflected in an extremely high VIF index and in a low Tolerance indicator (shown in Bold). A more detailed insight is provided in the correlation table obtained from Pearson analysis, as shown in Table 13.

Table 13. Pearson Correlation parameters from the *Energy* Use Intensity group.

	IESRI	Energy_Consumption per Capita	Energy_Intensity	Petroleum_Intensity
IESRI	1.000	−0.516	0.080	0.035
Energy_Consumption per Capita	0.516	1.000	−0.578	−0.543
Energy_Intensity	0.080	−0.578	1.000	**0.995**
Petroleum_Intensity	0.035	−0.543	**0.995**	1.000

Correlation between the Energy Intensity and Petroleum Intensity parameters, which is almost 100%, unequivocally indicates that they do not individually contribute to the explanation of IESRI as a dependent variable.

3.6. Electric Power Sector

Electric Power Sector is a group of variables that do not contribute significantly (48.7%) to the formation of IESRI values, as defined by multiple regression and shown in Table 14.

Table 14. The result of Multiple Regression Analysis for parameters from the Electric Power Sector group.

Model	R	R Square	Adjusted R Square	Std. Error Estimate
1	0.698	**0.487**	0.457	55.82863

The multicollinearity test identified the only two parameters from this category (Electricity Diversity and Non-Carbon Generation), and these parameters statistically significantly (Sig, shown in Bold) contribute to the IESRI explanation, as shown in Table 15.

Table 15. Multicollinearity and partial correlation of parameters from the Electric Power Sector group.

Parameter	Beta	T test	Sig	Partial	Tolerance	VIF
(Constant)		−1.808	0.080			
Electricity_Diversity	−0.614	−3.191	0.003	−0.480	0.407	2.456
Non-CO$_2$_Emitting_Share_of_Electricity_Generation	1.050	5.457	0.000	0.683	0.407	2.456

At the same time, the Electricity Diversity and Non-Carbon Electricity Generation parameters meet the condition of mutual independence (values of Tolerance > 0.1 and VIF values < 10).

3.7. Transportation Sector

Transportation Sector is a group of variables that, as in the case of the Energy Use Intensity group, record the smallest contribution (40.2%, shown in Bold) to the formation of IESRI values Table 16.

Table 16. The result of Multiple Regression Analysis for parameters from the Transportation Sector group.

Model	R	R Square	Adjusted R Square	Std. Error Estimate
1	0.634	**0.402**	0.367	60.26841

The only two parameters from this category: Transportation Energy per Capita and Transportation Energy Intensity, statistically significantly (Sig value, shown in Bold) contribute to the IESRI explanation Table 17.

Table 17. Multicollinearity and partial correlation of parameters from the Transportation Sector group.

Parameter	Beta	T test	Sig	Partial	Tolerance	VIF
(Constant)		8.327	0.000			
Transportation Energy per Capita	**−0.805**	**−4.718**	**0.000**	**−0.629**	**0.604**	**1.654**
Transportation Energy Intensity	**−0.610**	**−3.579**	**0.001**	**−0.523**	**0.604**	**1.654**

At the same time, these two parameters from the Transportation Sector category meet the condition of mutual independence (Tolerance values greater than 0.1 and VIF less than 10).

3.8. Environmental

Parameters for the Environmental category contribute 58.4% (shown in Bold) to the formation of the total IESRI Table 18.

Table 18. The result of Multiple Regression Analysis for parameters from the Environmental group.

Model	R	R Square	Adjusted R Square	Std. Error Estimate
1	0.764	**0.584**	0.547	51.00798

All individual parameters within this category significantly (Sig << 0.05, shown in Bold) contribute to the determination of IESRI, and the problem of multicollinearity exists in each of them (extremely high VIF indicator), as shown in detail in Table 19.

Table 19. Multicollinearity and partial correlation of parameters from the Environmental group.

Parameter	Beta	T test	Sig	Partial	Tolerance	VIF
(Constant)		2.550	0.016			
CO2_Emissions_Trend	5.553	5.005	**0.000**	0.657	0.010	**97.750**
CO2_ Emissions per Capita	−4.474	−5.637	**0.000**	−0.700	0.020	**50.029**
CO2_GDP_Intensity	2.679	4.599	**0.000**	0.625	0.037	**26.957**

Pearson correlation test shows the existence of a significant positive correlation between the CO_2 Emissions Trend and CO_2 Emissions per Capita parameters, as well as a high negative correlation between the CO_2 Emissions Trend and CO_2 GDP Intensity parameters (shown in Bold). The results of this test are shown in Table 20.

Table 20. Pearson Correlation parameters from the Energy Use Intensity group.

	IESRI	CO_2_Emissions_Trend	CO_2_ Emissions per Capita	CO_2_GDP_Intensity
IESRI	1.000	−0.286	−0.491	−0.008
CO_2_Emissions_Trend	−0.286	1.000	**0.866**	**−0.733**
CO_2_ Emissions per Capita	−0.491	**0.866**	1.000	−0.309
CO_2_GDP_Intensity	−0.008	**−0.733**	−0.309	1.000

The correlation between the the CO_2 Emissions Trend and CO_2 Emissions per Capita variables imposes the conclusion that there is no need to use them together, as this impairs the methodological correctness of the model.

4. Discussion and Recommendations

The results of detailed analyses of the variables included in the IESRI calculation indicate weaknesses of the methodology used. The authors sought to create a model whose complexity is also its greatest weakness. The essential problem lies in a large number of variables (29) and in a way in which the variables are distributed by groups. The variables were grouped in order to apply common metrics to them. The importance of variables is determined by weighting values assigned to them. Thus, the importance of a group is expressed as the sum of weighting values of its variables. Given that the sum of all weights in the IESRI calculation model is 100%, it is concluded that the authors used an empirical model to express the significance of groups and variables. To prove this claim, the regression analysis of variables by groups (as independent variables) was used in the research in order to express the importance of each group for the formation of IESRI (as dependent variable).

The obtained R squared values were used to express the significance of individual categories (Section 3.1, Section 3.2, Section 3.3, Section 3.4, Section 3.5, Section 3.6, Section 3.7, Section 3.8). If normalized, these values could be used for a different distribution of weighting values by categories Table 21.

Table 21. IESRI vs proposed weight distribution.

Model	Global Fuel	Fuel Import	Energy Consumptions	Price & Market Volatility	Energy use Intensity	Electric Power Sector	Transportation Sector	Environmental
IESRI model	14	17	20	15	14	7	7	6
Proposed model	16.6	17.5	15.8	14.7	7.5	9.2	7.5	10.9

Table 21 shows small differences in some groups of variables (Fuel import, Price & Market Volatility and Transportation), but significant differences in most categories. The defined IESRI model gives too much importance to the Energy consumptions and Energy use intensity categories. The model proposed by the authors shows a more uniform distribution of weighting values by categories (the standard deviation of the proposed model is 4.15 compared to the IESRI model where it is 5.21). It emphasizes the importance of the Environmental category, and significantly reduces the importance of the Energy consumptions and Energy use intensity categories, giving added importance to the Global fuel, Fuel import and Electric power sector categories.

Another, no less significant result is the assessment of mutual correlativity of the parameters for verifying the correctness of the method of obtaining IESRI. Multicollinearity test is used as an instrument, and the obtained result at the IESRI level shows that none of the variables meet the VIF criterion < 10. In order to decompose the problem, multicollinearity test was performed at the category level.

The obtained results indicate that only the Electric power sector and Transportation sector groups meet the determination and correlativity conditions. The correlativity condition indicates that individual variables from the same group should be combined, i.e., that two or more variables should be replaced by one variable. For example, the Energy Intensity and Petroleum Intensity variables in the Energy Use Intensity group are problematic due to their small impact on the formation of IESRI. The inflation of value of these two variables was due to cross-correlation at almost 100%. On the other hand, there are variables whose individual importance for the formation of IESRI is insignificant, although they do not correlate with other parameters from their group. For example, the GDP per Capita parameter from the Price & Market Volatility group is far below the significance limit and represents a candidate for exclusion from the model.

The results of this study indicate the general and the particular problems that exist in the area of decision-making on the basis of data on the level of energy security and energy policy in general. The creators of different quantitative energy security indicators use different approaches and variables in their work, but they still relatively rarely check their reliability and robustness. In general, it is necessary and recommended that, when creating and using the index for measuring energy security, the concept of energy security (for which plans and assessments are made) has to be defined, because it has a different essence in different countries, as well in the same country, but in different time periods. Priorities are certainly changing, which could be followed by the creation of dynamic indices, taking into account the observed specifics, which is often not the case at the moment. Curently, in most cases, the level of energy security is determined on the basis of a methodology that is defined in advance and equal to all, countries are ranked and receive recommendations based on that, very important decisions often are made on the basis of these assessments.

Second, it is necessary to select the variables that will be used to measure energy security. There are a large number of basic, but even more complex variables (as in the case of the index observed in this paper), which are not adequately selected to represent a particular phenomenon, or, which is methodologically absolutely unacceptable, are contained in each other (high degree of overlap observed). Furthermore, it was observed that traditional energy security indicators (related to security of supply and energy prices) still have the greatest weight and significance in describing overall energy security. The inclusion of additional variables is certainly desirable, especially indicators that describe the environmental impact, but the results obtained could be unrealistic and misleading. For example, the indicators of harmful gas emissions (in combination with others) can by no means be attributed solely and exclusively to the use of energy sources. In addition, the use of GPD and GDP related variables is further quesionable, as there are a significant number of criticisms of this macroeconomic indicator, but for now there is no clear replacement.

Due to all the above, it is necessary to re-examine the type and number of variables and it is not necessary to methodologically complicate the assessment (which leads to erroneous results) by including additional indices at all costs. Certainly, energy transition towards more sustainable path is certainly imposed by the need to incorporate new indicators, but their uncontrolled inclusion and neglect of reliable indicators that have been used for decades, does not bring more correct and more accurate results. On the contrary, exaggeration in the number of indicators is both fundamentally and methodologically wrong. The number of variables is nowhere clearly defined, but they can be included gradually, on a previously defined set of indicators that are methodologically reliable, and after the inclusion of each new variable, a reliability check should be done. This iterative procedure allows to immediately include or exclude variables for which a certain reliability assessment is made. Certainly, grouping variables makes sense (as defined by the observed Index), but the groups showed clear differences, with overlaps, so that certain indicators should not be part of the group in which they are classified, or they should be excluded in total. It can be assumed that the grouping of variables and their further analysis within groups (especially the relationship between them) is expedient and can yield significant results in a methodological sense.

The choice of data processing method is a special issue. The research presented in this paper provides an insight into the analysis of the reliabiliyty testing that was applied on the structure of the index itself, which should be applied (and certainly extended) to each index. Therefore, every evaluation of energy security level should be followed by both reliability and robustness check of the methodology used, because energy security is too important issue to be assessed by using methodologically incorrect and questionable procedures.

5. Conclusions

The main goal of this study is to assess the reliability of the International Energy Security Risk Index as an energy and economy transition decision making tool. Transition takes place under conditions of high uncertainty, so decision-making gains increasing importance. If we take into account that said Index refers to energy security and risk indicators, connected with economic and environmental indicators, it is highly necessary to establish whether said Index is sufficiently reliable, whether there are errors in its settings and how they could be corrected.

The research included the data for 25 countries in the period from 1980 to 2016, using Multiple Regression Analysis, Multicollinearity test plus Multicollinearity tests with Variance Inflation Factors and Pearson correlation tests. The study analyzed the impact of 29 variables on the total IESRI variance and the degree of multicollinearity between the variables, both as an entirety and as 8 groups to which the variables belong.

The research showed that the most reliable indicators, which do not show a critical degree of multicollinearity, are variables from the Fuel imports group: Oil Import Exposure, Gas Import Exposure, Coal Import Exposure, Fossil Fuel Import Expenditure per GDP and Total Energy Import Exposure. These variables are indicators of import exposure in general, which indicates great significance of the dependence on energy imports and methodological correctness of the settings of each of these variables. If the number of variables that make up the Index is reduced in further research (in order to increase its consistency and reliability), these variables are certainly to remain part of the Index. On the other hand, a high correlation measured for the Energy Intensity and Petroleum Intensity variables indicates that they insignificantly contribute to IESRI and that their further contribution needs to be reconsidered. Moreover, although the variables from the Transportation sector group (Transportation Energy per Capita and Transportation Energy Intensity) have the smallest contribution to the IESRI variance, both show a high degree of independence and a relatively small degree of overlap, so they should certainly remain part of IESRI.

The analysis of correlation of the economic group of variables showed a high degree of multicollinearity between GDP-related and macroeconomic indicators and pointed to the fact that their use should be further considered in more detail, while some should be excluded. The results show that the Energy Expenditures per Capita and Retail Electricity Prices variables have a negligibly small impact on the formation of IESRI, while the impact of the Crude Oil Price Volatility and Energy Expenditure Volatility variables is much greater.

In summary, the biggest problems are seen in the methodological settings of International Energy Security Risk Index. Specifically, independent variables should be independent in their essence, which is not the case here. On the contrary, all 29 variables show unacceptably high degrees of multicollinearity (>5), that consequently lead to inaccurate results that, accordingly, represent poor output for decision making. A high degree of multicollinearity occurs when weighting coefficients are set incorrectly, while the obtained p-values are certainly misleading. Due to the above, the authors propose a new redistribution of weighting coefficients in accordance with the results of this research.

Based on the presented results, some further research directions can be suggested. There should be a focus on reengineering of this index and on reviewing other globally used indices, in order to obtain reliable indices that can be valuable tools for decision makers. The emphasis should be, above all, on solving the problem of the selection of variables. The extent to which the selection is necessary and the potential overlap with the existing variables should be taken into consideration. Furthermore, when including (or excluding) each variable it should be checked whether changes have occurred in the reliability of the Index. The ultimate goal is to develop an index model that will contain variables relevant to the observed phenomenon and independent (the degree of multicollinearity within the limits allowed), while striving to respect the parsimony principle, i.e., to use only as many variables as really necessary (principle of simplicity), but also to use all those variables that are really necessary (principle of optimality). Modern trends and efforts to include macroeconomic and environmental indicators in the assessment of energy security, in addition to indicators on energy resources and imports, basically represent a sustainable energy transition. However, in this case it has been proven that the selection of variables is highly inaccurate, so it is certainly necessary to conduct a comprehensive audit. Finally, due to a high degree of uncertainty in the field of decision making for energy security and energy transition in general, it is recommended not to use only mathematical models, but also to examine the possibility of using fuzzy logic and other systems for inexact reasoning.

Author Contributions: Conceptualization, I.P., M.R., S.F., G.Š., D.M. and P.Š.; methodology, M.R., S.F. and G.Š.; validation, I.P., M.R., S.F., G.Š., D.M. and P.Š.; formal analysis, G.Š. and M.R.; investigation, I.P., M.R. D.M. and P.Š.; resources, D.M.; data curation, G.Š. and M.R.; writing—original draft preparation, I.P., G.Š., M.R., S.F., D.M. and P.Š.; writing—review and editing, I.P., G.Š., M.R., S.F., D.M. and P.Š.; supervision, M.R., S.F. and P.Š.; All authors have read and agreed to the published version of the manuscript.

Funding: This research received no external funding.

Acknowledgments: The second author was partially supported by Ministry of Science, Technological Development and Education Republic of Serbia (grant No. MTR 44007 III). The fifth author was supported by the Slovenian Research Agency; program No. P5-0018-Decision Support Systems in Digital Business. Co-funded by the Erasmus+ Programme of the European Union .

References

1. Filipović, S.; Radovanović, M.; Golušin, V. Macroeconomic and political aspects of energy security–Exploratory data analysis. *Renew. Sustain. Energy Rev.* **2018**, *97*, 428–435. [CrossRef]

2.   Soroudi, A.; Amraee, T. Decision making under uncertainty in energy systems: State of the art. *Renew. Sustain. Energy Rev.* **2013**, *28*, 376–384. [CrossRef]

3.   Maier, H.R.; Guillaume, J.; Van Delden, H.; Riddell, G.; Haasnoot, M.; Kwakkel, J.H. An uncertain future, deep uncertainty, scenarios, robustness and adaptation: How do they fit together? *Environ. Model. Softw.* **2016**, *81*, 154–164. [CrossRef]

4.   Podbregar, I.; Šimić, G.; Radovanović, M.; Filipović, S.; Šprajc, P. International energy security risk index—Analysis of the methodological settings. *Energies* **2020**, *13*, 3234. [CrossRef]

5.   Bhave, A.G.; Conway, D.; Dessai, S.; Stainforth, D.A. Barriers and opportunities for robust decision making approaches to support climate change adaptation in the developing world. *Clim. Risk Manag.* **2016**, *14*, 1–10. [CrossRef]

6.   Vinogradova, I. Multi-attribute decision-making methods as a part of mathematical optimization. *Mathematics* **2019**, *7*, 915. [CrossRef]

7.   Le, T.H.; Nguyen, C.P. Is energy security a driver for economic growth? Evidence from a global sample. *Energy Policy* **2019**, *129*, 436–451. [CrossRef]

8.   Lutz, C.; Lindenberger, D.; Kemmler, A. Macroeconomic effects of the energy transition. *Final Rep. Summ. Proj.* **2014**, *31*, 13.

9.   Stavytskyy, A.; Kharlamova, G.; Giedraitis, V.; Šumskis, V. Estimating the interrelation between energy security and macroeconomic factors in European countries. *J. Int. Stud.* **2018**, *11*, 217–238. [CrossRef]

10.   Šumskis, V.; Giedraitis, V. Economic implications of energy security in the short run. *EÈkonomika* **2015**, *94*, 119–138. [CrossRef]

11.   Punzi, M.T. The impact of energy price uncertainty on macroeconomic variables. *Energy Policy* **2019**, *129*, 1306–1319. [CrossRef]

12.   Bergmann, P. Oil price shocks and GDP growth: Do energy shares amplify causal effects? *Energy Econ.* **2019**, *80*, 1010–1040. [CrossRef]

13.   Ju, K.; Su, B.; Zhou, D.; Wu, J.; Liu, L. Macroeconomic performance of oil price shocks: Outlier evidence from nineteen major oil-related countries/regions. *Energy Econ.* **2016**, *60*, 325–332. [CrossRef]

14.   Fotis, P.; Karkalakos, S.; Asteriou, D. The relationship between energy demand and real GDP growth rate: The role of price asymmetries and spatial externalities within 34 countries across the globe. *Energy Econ.* **2017**, *66*, 69–84. [CrossRef]

15.   Andrei, J.V.; Mieila, M.; Panait, M. The impact and determinants of the energy paradigm on economic growth in European Union. *PLoS ONE* **2017**, *12*, e0173282. [CrossRef]

16.   Sa, A.; Thollander, P.; Rafiee, M. Industrial energy management systems and energy-related decision-making. *Energies* **2018**, *11*, 2784. [CrossRef]

17.   Narula, K.; Reddy, B.S. A SES (sustainable energy security) index for developing countries. *Energy* **2016**, *94*, 326–343. [CrossRef]

18.   Razmjoo, A.A.; Sumper, A.; Davarpanah, A. Energy sustainability analysis based on SDGs for developing countries. *Energy Sources Part A Recover. Util. Environ. Eff.* **2019**, *42*, 1041–1056. [CrossRef]

19.   Siksnelyte, I.; Zavadskas, E.K.; Štreimikienė, D.; Sharma, D. An Overview of multi-criteria decision-making methods in dealing with sustainable energy development issues. *Energies* **2018**, *11*, 2754. [CrossRef]

20.   Pérez, M.D.L.E.M.; Scholten, D.; Stegen, K.S. The multi-speed energy transition in Europe: Opportunities and challenges for EU energy security. *Energy Strat. Rev.* **2019**, *26*, 100415. [CrossRef]

21.   Radovanović, M.; Filipović, S.; Golušin, V. Geo-economic approach to energy security measurement–principal component analysis. *Renew. Sustain. Energy Rev.* **2018**, *82*, 1691–1700. [CrossRef]

22.   Radovanovic, M.; Filipović, S.; Pavlović, D. Energy security measurement–A sustainable approach. *Renew. Sustain. Energy Rev.* **2017**, *68*, 1020–1032. [CrossRef]

23.   Månsson, A.; Johansson, B.; Nilsson, L.J. Assessing energy security: An overview of commonly used methodologies. *Energy* **2014**, *73*, 1–14. [CrossRef]

24.   Sovacool, B.K.; Mukherjee, I. Conceptualizing and measuring energy security: A synthesized approach. *Energy* **2011**, *36*, 5343–5355. [CrossRef]

25.   Fang, D.; Shi, S.; Yu, Q. Evaluation of sustainable energy security and an empirical analysis of China. *Sustainability* **2018**, *10*, 1–23. [CrossRef]

26.   European Commission. *Towards a European Strategy for the Security of Energy Supply*; Office for Official Publications of the European Communities: Luxembourg, 2000.

27. Cherp, A.; Jewell, J. The concept of energy security: Beyond the four As. *Energy Policy* **2014**, *75*, 415–421. [CrossRef]

28. Erahman, Q.F.; Purwanto, W.W.; Sudibandriyo, M.; Hidayatno, A. An assessment of Indonesia's energy security index and comparison with seventy countries. *Energy* **2016**, *111*, 364–376. [CrossRef]

29. Šprajc, P.; Bjegović, M.; Vasić, B. Energy security in decision making and governance–Methodological analysis of energy trilemma index. *Renew. Sustain. Energy Rev.* **2019**, *114*, 109341. [CrossRef]

30. Li, Y.; Shi, X.; Yao, L. Evaluating energy security of resource-poor economies: A modified principle component analysis approach. *Energy Econ.* **2016**, *58*, 211–221. [CrossRef]

31. Hong, J.H.; Kim, J.; Son, W.; Shin, H.; Kim, N.; Lee, W.K.; Kim, J. Long-term energy strategy scenarios for South Korea: Transition to a sustainable energy system. *Energy Policy* **2019**, *127*, 425–437. [CrossRef]

32. Brown, M.; Wang, Y.; Sovacool, B.K.; D'Agostino, A. Forty years of energy security trends: A comparative assessment of 22 industrialized countries. *Energy Res. Soc. Sci.* **2014**, *4*, 64–77. [CrossRef]

33. Augutis, J.; Krikštolaitis, R.; Martišauskas, L.; Urbonienė, S.; Urbonas, R.; Ušpurienė, A.B. Analysis of energy security level in the Baltic States based on indicator approach. *Energy* **2020**, *199*, 117427. [CrossRef]

34. Martišauskas, L.; Augutis, J.; Krikštolaitis, R. Methodology for energy security assessment considering energy system resilience to disruptions. *Energy Strat. Rev.* **2018**, *22*, 106–118. [CrossRef]

35. Azzuni, A.; Breyer, C. Definitions and dimensions of energy security: A literature review. *Wiley Interdiscip. Rev. Energy Environ.* **2017**, *7*, e268. [CrossRef]

36. Ang, B.; Choong, W.; Ng, A. Energy security: Definitions, dimensions and indexes. *Renew. Sustain. Energy Rev.* **2015**, *42*, 1077–1093. [CrossRef]

37. Gasser, P. A review on energy security indices to compare country performances. *Energy Policy* **2020**, *139*, 111339. [CrossRef]

38. *International Index of Energy Security Risk*; US Chamber of Commerce: Washington, DC, USA, 2018. Available online: https://www.globalenergyinstitute.org/sites/default/files/energyrisk_intl_2018.pdf (accessed on 20 May 2020).

39. Takahashi, S.; Inoue, I. *The Manga Guide to Regression Analysis*; No Starch Press: San Francisco, CA, USA, 2016.

40. Daoud, J.I. Multicollinearity and Regression Analysis. *J. Physics: Conf. Ser.* **2017**, *949*, 012009. [CrossRef]

Review

Internet of Things (IoT) and the Energy Sector

Naser Hossein Motlagh [1], Mahsa Mohammadrezaei [2] and Julian Hunt [3]
and Behnam Zakeri [3,4,*]

1. Department of Computer Science, University of Helsinki, FI-00560 Helsinki, Finland; naser.motlagh@helsinki.fi
2. Telfer School of Management, University of Ottawa, Ottawa , ON K1N 6N5, Canada; mmoha296@uottawa.ca
3. International Institute for Applied Systems Analysis (IIASA), A-2361 Laxenburg, Austria; hunt@iiasa.ac.at
4. Sustainable Energy Planning, Department of Planning, Aalborg University, 2450 Copenhagen, Denmark
* Correspondence: zakeri@iiasa.ac.at

Received: 6 December 2019; Accepted: 13 January 2020; Published: 19 January 2020

Abstract: Integration of renewable energy and optimization of energy use are key enablers of sustainable energy transitions and mitigating climate change. Modern technologies such the Internet of Things (IoT) offer a wide number of applications in the energy sector, i.e, in energy supply, transmission and distribution, and demand. IoT can be employed for improving energy efficiency, increasing the share of renewable energy, and reducing environmental impacts of the energy use. This paper reviews the existing literature on the application of IoT in in energy systems, in general, and in the context of smart grids particularly. Furthermore, we discuss enabling technologies of IoT, including cloud computing and different platforms for data analysis. Furthermore, we review challenges of deploying IoT in the energy sector, including privacy and security, with some solutions to these challenges such as blockchain technology. This survey provides energy policy-makers, energy economists, and managers with an overview of the role of IoT in optimization of energy systems.

Keywords: internet of things; IoT applications; energy efficiency; energy in buildings; smart energy systems; smart grid; flexible demand; energy storage

---

1. Introduction

1.1. Concepts

Industrial revolutions can be divided into four phases. In the first revolution, new sources of energy were discovered to run the machines. The mass extraction of coal and the invention of steam power plants were significant development stages in this phase [1]. The second revolution known as mass production and electricity generation was a period of rapid development in industry, distinguished by large-scale iron and steel production. During this phase, many large-scale factories with their assembly lines were established and formed new businesses [2]. The third revolution introduced computer and the first generation of communication technologies, e.g., telephony system, which enabled automation in supply chains [3].

A wide variety of modern technologies such as communication systems (e.g., 5G), intelligent robots, and the Internet of Things (IoT) are expected to empower the fourth industrial revolution [4–6]. IoT interconnects a number of devices, people, data, and processes, by allowing them to communicate with each other seamlessly. Hence, IoT can help improving different processes to be more quantifiable and measurable by collecting and processing large amount of data [7]. IoT can potentially enhance the quality of life in different areas including medical services, smart cities, construction industry, agriculture, water management, and the energy sector [8]. This is enabled by providing an increased automated decision making in real-time and facilitating tools for optimizing such decisions.

1.2. Motivation

The global energy demand rose by 2.3% in 2018 compared to 2017, which is the highest increase since 2010 [9]. As a result, CO_2 emissions from the energy sector hit a new record in 2018. Compared to the pre-industrial temperature level, global warming is approaching 1.5 °C, most likely before the middle of the 21 Century [10]. If this trend prevails, the global warming will exceed the 2 °C target, which will have a severe impact on the planet and human life. The environmental concerns, such as global warming and local air pollution, scarcity of water resources for thermal power generation, and the limitation of depleting fossil energy resources, raise an urgent need for more efficient use of energy and the use of renewable energy sources (RESs). Different studies have shown that a non-fossil energy system is almost impossible without efficient use of energy and/or reduction of energy demand, and a high level integration of RESs, both at a country level [11], regional [12], or globally [13].

Based on the United Nations Sustainable Development Goals agenda [14], energy efficiency is one of the key drivers of sustainable development. Moreover, energy efficiency offers economic benefits in long-term by reducing the cost of fuel imports/supply, energy generation, and reducing emissions from the energy sector. For enhancing energy efficiency and a more optimal energy management, an effective analysis of the real-time data in the energy supply chain plays a key role [15]. The energy supply chain, from resource extraction to delivering it in a useful form to the end users, includes three major parts: (i) energy supply including upstream refinery processes; (ii) energy transformation processes including transmission and distribution (T&D) of energy carriers; and (iii) energy demand side, which includes the use of energy in buildings, transportation sector, and the industry [16]. Figure 1 shows these three parts with their relevant components. Under the scope of this paper, we discuss the role of IoT in all different segments of the energy supply chain. Our aim is to show the potential contribution of IoT to efficient use of energy, reduction of energy demand, and increasing the share of RESs.

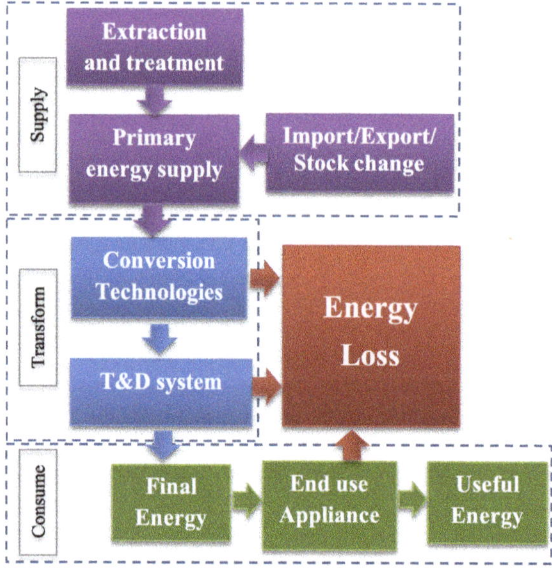

Figure 1. Energy supply chain.

IoT employs sensors and communication technologies for sensing and transmitting real-time data, which enables fast computations and optimal decision-making [17]. Moreover, IoT can help the energy sector to transform from a centralized to a distributed, smart, and integrated energy system. This is a key requirement in deploying local, distributed RESs, such as wind and solar energy [18], as well as turning many small-scale end users of energy into prosumers by aggregating their generation and optimizing their demand whenever useful for the grid. IoT-based systems automate, integrate, and control processes through sensors and communication technologies. Large data collection and use of intelligent algorithms for real-time data analysis can help to monitor energy consumption patterns of different users and devices in different time scales and control that consumption more efficiently [19].

1.3. Methodology

The application of IoT in different sectors and industries has been widely discussed and reviewed in the literature (for example [20–22]. Moreover, challenges and opportunities with respect to the deployment of one or a group of IoT technologies have received a high level of technical assessment, e.g., sensors [23] or 5G network [24]. With respect to the energy sector, most of survey studies have focused on one specific subsector, e.g., buildings or the technical potential of a certain IoT technology in the energy sector. For example, Stojkoska et al. [25] reviews smart home applications of IoT and the prospect of integrating those applications into an IoT enabled environment. In a study by Hui et al. [26], the methods, recent advances, and implementation of 5G are studied only with focusing on the energy demand side. The role of IoT in improving energy efficiency in buildings and public transport has been discussed in [27,28], respectively. Khatua et al. [29] reviews the key challenges in the suitability of IoT data transfer and communication protocols for deployment in smart grids.

However, unlike the reviewed literature where the focus is commonly either on a specific subsector in the energy sector or certain IoT technologies, this paper reviews the application of IoT in the energy sector, from energy generation to transmission and distribution (T&D) and demand side. As such, the main contribution of this paper is to extend the existing body of literature by providing energy policy-makers, economists, energy experts, and managers with a general overview of the opportunities and challenges of applying IoT in different parts of the entire energy sector. In this respect, we briefly introduce the IoT framework and its enabling technologies to form a basis for discussing their role in the energy sector.

To conduct this survey, a systematic search was carried out to collect and review the recent body of literature on the role of IoT in the energy sector. First, we searched the terms "Internet of Things" and "Energy", case non-specific, in the title, abstract, and keywords of publications stored in SCOPUS, IEEE, Hindawi, and Google Scholar databases. Then, we limited the scope of search results to engineering, economics and management branches where possible. Next, papers published before 2012 and most of conference papers with no information on the peer-review process were excluded. Finally, we clustered the relevant papers in sub-categories of energy generation (including power plants, ancillary services, and centralized renewable energy), T&D systems (including electricity, gas and district heating networks, and smart grids), and the demand side (including energy use in buildings, transportation, and the industry sector). We focus on the IoT applications that can be generally applicable to most of energy systems without discussing specific cases and their boundary conditions. For example, we discuss the role of IoT in smart buildings, without falling into the details of building typology, building material, occupants' energy consumption pattern, type and number of home appliances, etc.

The rest of this paper is structured as follows. Sections 2 and 3 introduces IoT and enabling technologies, including sensors and communication technologies, cloud computing, and data analytic platforms. Section 4 reviews the role of IoT in the energy sector. Section 5 discusses the opportunities and challenges of deploying IoT, while Section 6 portrays future trends. The paper concludes in Section 7.

2. Internet of Things (IoT)

IoT is an emerging technology that uses the Internet and aims to provide connectivity between physical devices or "things" [30]. Examples of physical devices include home appliances and industrial equipment. Using appropriate sensors and communication networks, these devices can provide valuable data and enable offering diverse services for people. For instance, controlling energy consumption of buildings in an smart fashion enables reducing the energy costs [31]. IoT has a wide range of applications, such as in manufacturing, logistics and construction industry [32]. IoT is also widely applied in environmental monitoring, healthcare systems and services, efficient management of energy in buildings, and drone-based services [33–36].

When planning an IoT application which is the first step in designing IoT systems, the selection of components of IoT such as sensor device, communication protocol, data storage and computation needs to be appropriate for the intended application. For example, an IoT platform planned to control heating, cooling, and air conditioning (HVAC) in a building, requires utilizing relevant environmental sensors and using suitable communication technology [37]. Figure 2 shows the different components of an IoT platform [38]. IoT devices which are the second components of the IoT platforms, could be in the form of sensors, actuators, IoT gateways or any device that joins the cycle of data collection, transmission, and processing. For example, an IoT gateway device enables routing the data into the IoT system and establishing bi-directional communications between the device-to-gateway and gateway-to-cloud.

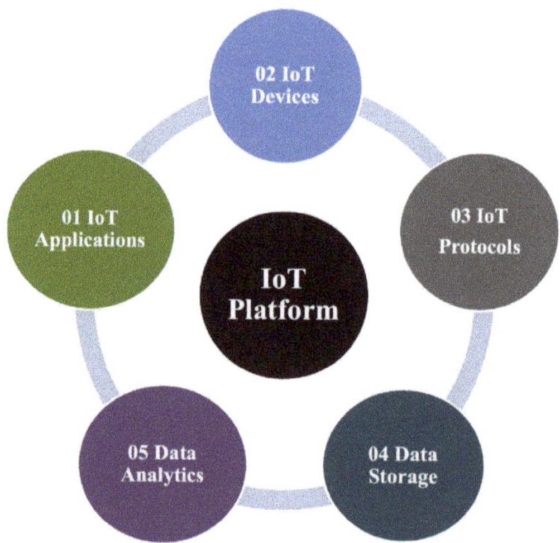

Figure 2. Diagram describing the components of an IoT platform.

The communication protocols that are the third component of the IoT platform, enable the different devices to communicate and share their data with the controllers or the decision making centers. IoT platforms offer the flexibility to select the type of the communication technologies (each having specific features), according to the needs of the application. The examples of these technologies include Wi-Fi, Bluetooth, ZigBee [39] and cellular technology such as LTE-4G and 5G networks [40]. The data storage is the fourth component of the IoT platform which enables management of collected data from the sensors.

In principle, the data collected from the devices is very large. This necessitates planning an efficient data storage that can be in cloud servers or at the edge of an IoT network. The stored data which is used for analytical purposes, forms the fifth component of the IoT platforms. The data analytics can be performed off-line after storing the data or it can be in form of real-time analytics. The data analytic is performed for decision making about the operation of the application. Based on the need, the data analytics can be performed off-line or real-time. In off-line analytics, the stored data is first collected and then visualized on premises using visualization tools. In case of real-time analytics, the cloud or edge servers are used to provide visualization, e.g. stream analytics [41].

3. Enabling Technologies

IoT is a paradigm in which objects and elements of a system that are equipped with sensors, actuators, and processors can communicate with each other to provide meaningful services. In IoT systems, sensors are used to sense and collect data, and through gateways route the collected data to control centers or the cloud for further storage, processing, analytics, and decision-making. After the decision is made, a corresponding command is then sent back to the actuator installed on the system in response to the sensed data. As there are variety of sensor and actuator devices, communication technologies, and data computing approached, in this section, we explain the existing technologies which enable IoT. Then, we provide examples from the literature how these technologies are used in the energy sector.

3.1. Sensor Devices

Sensors are the key drivers of IoT [42]. They are employed to collect and transmit data in real-time. The use of sensors enhances effectiveness, functionality, and plays a critical role in success of IoT [43]. Different types of sensors exist that are developed for various application purposes. The examples of these applications include agriculture industry, environmental monitoring, healthcare systems and services, and public safety [44]. In practice, in the energy sector including energy production, transmission and distribution, and production, many these sensors are used. In the energy sector, sensors are used to create savings in both cost and energy. Sensors enable smart energy management system and provide real-time energy optimization and facilitate new approaches for energy load management. The research and future trends of the sensor devices are also aims at development of sensor applications to improve load shaping and consumers' awareness as well as development of specific facilities to enhance production of renewable energies [45]. In nutshell, the use of sensor devices within IoT, in the energy sector largely improves diagnostics, decision-making, analytics, optimization processes and integrated performance metrics. Due to the large number of sensors used in the energy sector, in the following we explain few examples of commonly applied sensor devices in energy production and consumption.

Temperature sensors are used to detect the fluctuations in heating and cooling a system [46]. Temperature is an important and common environmental parameter. In the energy sector, the basic principle of power generation is the process of changing mechanical energy into electrical energy, whereas mechanical energy is achieved from heat energy, e.g., thermal power plants, wind, water flow, and solar power plants. These energy conversions are obtained using thermal, i.e., temperature. In the energy consumption side, the temperature sensors are used to maximize the performance of a system when temperature changes during normal operations. For example, in residential areas the best time for turning on or off the ventilation and cooling systems is recognized by temperature sensors; thus, the energy can be managed correctly in order to save energy [42].

Humidity sensors are used to distinguish the amount moisture and air's humidity. The ratio of moisture in the air to the highest amount of moisture at a particular air temperature is called relative humidity [47]. The applications of humidity sensors in the energy sector are wide. For example, they are generally used in wind energy production. The use of humidity sensors on wind turbines are even vital, if the turbines are located offshore (due to the high level of moisture in the air). Humidity sensors can be placed in the nacelle and in the bottom of wind turbines and offer continuous moisture monitoring. This enables the operators to take actions to changes or deviations in the turbine operation conditions, leading to more consistent operations, optimized performance, and lower costs of energy.

Light sensors are used to measure luminance (ambient light level) or the brightness of a light. In energy consumption, light sensors have several uses in industrial and everyday consumer applications. As a main source of energy consumption in buildings relates to lighting, which, respectively, account for nearly 15% of total electricity usage [48]. On global scale, approximately 20% of electricity is consumed for lighting [49]. Therefore, light sensors can be utilized to automatically control lighting levels indoors and outdoors by turning on-and-off or dimming the light levels, such that the electric light levels automatically can be adjusted in response to changes in ambient light. In this fashion, the energy required for the lighting for the indoor environments can be reduced [19].

Passive Infrared (PIR) sensors, also known as motion sensors, are used for measuring the infrared light radiation emitted from objects in their surroundings. In energy consumption, these sensors are used to reduce the energy consumption in buildings. For example, by using PIR sensors, the presence of humans inside spaces can be detected. If there is no movement detected in the space then the light control of the space turns off the light, i.e., smart control of lighting. In this fashion, the electricity consumption of the buildings are reduced [50]. Similarly, this can be applied for air conditioning systems which consume nearly 40% of the energy in buildings [48].

Proximity sensors are utilized to detect the presence of nearby objects without any physical contact [51]. The example application of proximity sensors is in wind energy production. These sensors provide longevity and reliable position sensing performance in wind turbines. In wind turbines, the applications of proximity sensors include blade pitch control, yaw position, rotor, and yaw brake position; brake wear monitoring; and rotor speed monitoring [52].

3.2. Actuators

Actuators are devices that transform a certain form of energy into motion. They take electrical input from the automation systems, transform the input into action, and act on the devices and machines within the IoT systems [53]. Actuators produce different motion patterns such as linear, oscillatory, or rotating motions. Based on the energy sources, actuators categorized as the following types [54].

Pneumatic actuators use compressed air for generating motion. Pneumatic actuators are composed of a piston or a diaphragm in order to generate the motive power. These actuators are used to control processes which require quick and accurate response, as these processes do not need a large amount of motive force.

Hydraulic actuators utilize the liquid for generating motion. Hydraulic actuators consist of cylinder or fluid motor that uses hydraulic power to provide mechanical operation. The mechanical motion gives an output in terms of linear, rotatory, or oscillatory motion. These actuators are used in industrial process control where high speed and large forces are required.

Thermal actuators use a heat source for generating the physical action. Thermal actuators convert thermal energy into kinetic energy, or motion. Generally, thermostatic actuators are composed of a temperature-sensing material sealed by a diaphragm which pushes against a plug for moving a piston. The temperature-sensing material can be any type of liquid, gas, wax-like substance, or any material that changes volume based on temperature.

Electric actuators apply external energy sources, e.g., batteries to generate motion. Electric actuators are mechanical devices capable of converting electricity into kinetic energy in either a single linear or rotary motion. The designs of these actuators are based on the intended tasks within the processes.

In the energy sector, for example, in power plants, pneumatic actuators are traditionally applied to control valves. Electric control-valve actuator technology enables achieving energy efficiency. They are also often used as the final control element in the operation of a power plant [55]. In addition, there are variety of actuators developed for energy industry, e.g., LINAK electric actuator (https://www.linak.com/business-areas/energy/) that offer solutions for example for minimizing the energy waste when opening hatches and locking brakes in wind turbines and creating motion in solar tracking panels. In the literature there are also many studies aimed to illustrate the applications of the actuators within IoT. For instance, the research in [56] proposes a wireless sensor and actuator network to provide an IoT-based automatic intelligent system. Whereas, by optimizing the operation of devices and machines in the IoT, the proposed system achieves reduction in their overall energy consumption at a given time.

3.3. Communication Technologies

Wireless communication systems play the major role in activating IoT. Wireless systems connect the sensor devices to IoT gateways and perform end-to-end data communications between these elements of IoT. Wireless systems are developed based on different wireless standards and the use of each one depends on the requirement of the application such as communication range, bandwidth, and power consumption requirements. For example, often renewable sources of energy, including wind and solar power plants are mostly located in very remote areas. Therefore, ensuring a reliable IoT communications in remote places is challenging. Employing IoT systems on these sites requires selection of suitable communication technology that can guarantee a continues connection link and support real-time data transfer in an energy efficient manner. Due to the importance of communication technologies in IoT, in this subsection we review some of these technologies. We also indicate to few examples to show their role in the energy sector. Then, we provide a comparison in Table 1 to show the difference of each of the technologies when applied with IoT.

The short range wireless technologies, e.g., Wireless Fidelity (Wi-Fi) (https://www.wi-fi.org/) for IoT applications in the energy sector has been widely studies. In the energy sector, the obvious cases of using Wi-Fi include energy metering and building energy management [57–61]. However, due to high power requirements of Wi-Fi, this technology is not the best solution in the energy sector. Low power wide area network (LPWAN) communication technologies such as narrowband IoT (NB-IoT); ZigBee; Bluetooth low energy (BLE) technologies; as well as the emerging LPWAN technologies such as LoRa, Sigfox, and LTE-M operating in unlicensed band [62] are better solutions to be used in the energy sector. Because, these emerging LPWAN technologies enable establishing a reliable, low-cost, low-power, long-range, last-mile technology for smart energy management solutions [63]. Therefore, in this paper, we explain the short range and emerging LPWAN technologies and review some examples for their applied cases in the energy sector. We also explain the satellite technology which plays an important role in providing global IoT connectivity for industrial sectors located remote areas. In addition, in Table 1, we illustrate the different features of these technologies.

Bluetooth Low Energy (BLE) is a short range wireless communication technology for IoT that enables exchanging data using short radio wavelengths (https://www.bluetooth.com/). BLE is less costly to deploy, with a typical range of 0 to 30 m, which enables creating an instant personal area network [64]. BLE targets small-scale IoT applications that require devices to communicate small volumes of data consuming minimal power. Industries in the energy sector with a well-designed IoT strategy can create new forms of machine-to-machine and machine-to-human communication using this technology. In the energy sector, BLE is widely used on the energy consumption in residential and commercial buildings. For instance, the authors of [65] describe a smart office energy management system that reduces the energy consumption of PCs, monitors, and lights using BLE. Another study proposes an energy management system for smart homes that utilizes BLE for communication among home appliances aiming at decreasing the energy at homes [66]. Similarly, using BLE the research in [67] introduces a fuzzy-based solution for smart energy management in a home automation, aiming improving home energy management scheme.

Zigbee is a communication technology, which is designed to create personal area network and targets small scale applications (https://zigbee.org/). Zigbee is easy to implement and planned to provide low-cost, low-data rate, and highly reliable networks for low-power applications [68,69]. Zigbee also utilizes a mesh network specification where devices are connected with many interconnections. Using the mesh networking feature of Zigbee, the maximum communication range, which is up to 100 m, is extended significantly. In the energy sector, the example IoT applications of Zigbee include lighting systems (buildings and street lighting), smart grids, e.g., smart electric meters, home automation systems and industrial automation. These applications aim to provide approaches for consuming energy in an efficient way. In literature, aiming to minimize the energy expenses of the consumers, the research in [70] evaluates the performance of home energy management application through establishing a wireless sensor network using Zigbee. The authors of [71] also introduce smart home interfaces to allow interoperability among ZigBee devices, electrical equipment, and smart meters to utilize the energy more efficiently. The work in [72] presents a ZigBee-based monitoring system which is used to measure and transfer the energy of home appliances at the outlets and the lights, aiming at reducing the energy consumption. Another study [73] presents field tests using ZigBee technologies for monitoring photovoltaic and wind energy systems. The results of the study demonstrate the proficiency of ZigBee devices applied in distributed renewable generation and smart metering systems.

Long Range (LoRa) LoRa is a wireless communication technology designed for IoT (https://lora-alliance.org/). LoRA is a cost-effective communication technology for large deployment of IoT, it can add many years to battery life. LoRa is also used to establish long-distance broadcasts (more than 10 km in rural areas) with very low power consumption [74]. The features of this technology make it a suitable communication technology to be used in the energy sector mainly in smart cities, such as smart grids and building automation systems, e.g., smart metering.

In literature, the work in [75] aims at optimizing energy consumption by deploying building energy management system using LoRa. The work proposes a platform by integrating multiple systems, such as air-conditioning, lighting, and energy monitoring to perform building energy optimization. The outcome of platform resulted in a 20% energy saving. In [76], the authors developed a machine-learning-based smart controller for a commercial buildings HVAC utilizing LoRa. The smart controller identifies when a room is not unoccupied and turns off the HVAC, reducing its energy consumption up to 19.8%. Using LoRa technology, another study [77] presents implementation of an electronic platform for energy management in public buildings. Through a test, the developed platform allows saving the energy for a lighting system by 40%.

Sigfox is a wide area network technology which uses an ultra narrow band (https://www.sigfox.com/). Sigfox allows devices to communicate with low power for enabling IoT applications [78]. For the appropriateness of this technology in the energy sector for example, the study in [79] reviews the technological advances and presents Sigfox as one of the best low power candidates for smart metering for enabling real-time energy services for households. In addition, the study in [80] compare different low power wide area network technologies and conclude that Sigfox is a suitable solution to be used with electric plugs sensors alert in smart buildings.

Narrowband IoT (NB-IoT) is a LPWAN communication technology that supports large number of IoT devices and services with a high data rate with very low latency (https://www.3gpp.org/news-events/1733-niot/). NB-IoT is a low-cost solution that has long battery life and provides enhanced coverage. According to the authors of [81], due to the latency features of NB-IoT, this technology is a potential solution for smart energy distribution networks by providing low-cost communications for smart meters. In addition, the study in [82] demonstrates the NB-IoT technology for smart metering. As another application of NB-IoT in the energy sector, the work in [83] introduces NB-IoT as a potential solution for smart grid communications by comparing NB-IoT with other communication technologies in terms of data rate, latency, and communication range.

Long Term Evolution for Machine-Type Communications (LTE-M) is the 3GPP (the third-generation partnership project) standardization, which is designed to reduce the device complexity for machine-type communication (MTC) [84]. LTE-M supports secure communication, provides ubiquitous coverage, and offers high system capacity. LTE-M also offers services of lower latency and higher throughput than NB-IoT [85]. In addition, this technology offers energy efficiency resource allocation for small powered devices, making it to be a potential solution for smart meter [86] and smart grid communications [87].

Weightless is LPWAN open wireless standard that is developed to establish communication among great number of IoT devices and machines (http://www.weightless.org/). Weightless is a potential solution for smart metering in the energy sector [88]. Based on the study in [89], Weightless is a suitable wireless technology can be used in smart home IoT applications for smart metering and smart grid communications.

Satellite is another communication technology that has a very wide-area coverage and can support low data rate applications in machine-to-machine (M2M) fashion [90]. Satellite technology is suitable for supporting IoT devices and machines in remote places. The study in [91] presents an IoT-based machine-to-machine satellite communication that is applicable to the smart grid, particularly for the transmission and distribution (T&D) sector. A similar study highlights the importance of using satellite-based IoT communications in energy domain such as solar and wind power plants [92].

Table 1. Comparison between different wireless technologies [62,93–98].

Technology \ Parameter	Range	Data Rate	Power Usage (Battery Life)	Security	Installation Cost	Example Application
LoRA	⩽50 km	0.3–38.4 kbps	Very low (8–10 years)	High	Low	Smart buildings (smart lighting)
NB-IoT	⩽50 km	⩽100 kbps	High (1–2 years)	High	Low	Smart grid communication
LTE-M	⩽200 km	0.2–1 Mbps	Low (7–8 years)	High	Moderate	Smart meter
Sigfox	⩽50 km	100 bps	Low (7–8 years)	High	Moderate	Smart buildings (electric plugs)
Weightless	<5 km	100 kbps	Low (Very Long)	High	Low	Smart meter
Bluetooth	⩽50 m	1 Mbps	Low (Few months)	High	Low	Smart home appliances
Zigbee	⩽100 m	250 Kbps	Very Low (5–10 years)	Low	Low	Smart metering in renewable energies
Satellite	Very Long >1500 km	100 kbps	High	High	Costly	Solar & wind power plants

3.4. IoT Data and Computing

Computing and analyzing the data generated by IoT allows gaining deeper insight, accurate response to the system, and helps making suitable decisions on energy consumption of the systems [99]. However, computing IoT data is a challenging issue. Because, IoT data known as Big data refers to huge amount of structured and unstructured data, generated from various elements of IoT systems such as sensors, software applications, smart or intelligent devices, and communication networks. Due to the characteristics of Big data, which are big volume, high velocity, and high variety [100], it needs to be efficiently processed and analyzed [101]. Processing the Big data is beyond the capacity of traditional methods, i.e., storing it on local hard drives, computing, and analyzing them afterwards. Advanced computing and analytic methods are needed to manage the Big data [102,103]. In the followings, we explain cloud computing and fog computing, which are widely used for processing and computing the Big data.

3.4.1. Cloud Computing

Cloud computing is a data processing approach that offers services, applications, storage, and computing through the internet and allows computation of data streamed from IoT devices. In cloud computing, cloud refers to the "Internet" and computing refers to computation and processing services offered by this approach [104]. Cloud computing consists of both application services that are accessed via the Internet and the hardware systems, which are located in data centers [105]. Using these characteristics, cloud computing enables processing the big data, and provides complex computation capabilities [106]. The main benefits of using cloud systems relies on [107] (i) significantly reducing the cost of hardware; (ii) enhancing the computing power and storage capacity; and (iii) having multi-core architectures, which eases the data management. Moreover, cloud computing is a secured system, which provides resources, computing power, and storage that is needed from a geographical location [108]. These features of cloud computing enables the big data resulted from the growing applications of IoT to be easily analyzed, controlled and sorted efficiently [109]. In addition, cloud computing eliminates the costs needed for purchasing hardware and software and running the algorithms for processing the IoT data, resulting in considerably minimization of electricity needed for local data computation.

3.4.2. Fog Computing

Although cloud computing is one of the best computing paradigms for data processing for IoT applications. Due to the delay and bandwidth limitation of centralized resources that are used for data processing, more efficient ways are required. Fog computing is a distributed paradigm and an extension of the cloud, which moves the computing and analytic services near to the edge of the network. Fog computing is a paradigm that expands the cloud at a greater scale and can support larger workload [110]. In fog computing, any device with computing, storage, and network connection capability works as fog node. The examples of these devices include, but are not limited to, personal computers, industrial controllers, switches, routers, and embedded servers [111]. In this computing paradigm, fog provides IoT data processing and storage locally at IoT devices instead of sending them to the cloud. The advantages of this approach include enhanced secure services required for many IoT applications as well as reducing network traffic and latency [112]. Therefore, in contrast to the cloud computing, fog offers processing and computing services with faster response with higher security. This enables faster decision-making and taking appropriate actions.

4. IoT in the Energy Sector

Today, the energy sector is highly dependent on fossil fuels, constituting nearly 80 % of final energy globally. Excessive extraction and combustion of fossil fuels has adverse environmental, health, and economic impacts due to air pollution and climate change to name a few. Energy efficiency, i.e., consuming less energy for delivering the same service, and the deployment of renewable energy sources are two main alternatives to diminish the adverse impacts of fossil fuel use [12,13].

In this section, we discuss the role of IoT in the energy sector, from fuel extraction, operation, and maintenance (O&M) of energy generating assets, to T&D and end use of energy IoT can play a crucial role in reducing energy losses and lowering CO_2 emissions. An energy management system based on IoT can monitor real-time energy consumption and increase the level of awareness about the energy performance at any level of the supply chain [15,113]. This section discusses the application of IoT in energy generation stages first. Then, we continue with the concept of smart cities, which is an umbrella term for many IoT-based subsystems such as smart grids, smart buildings, smart factories, and intelligent transportation. Next, we discuss each of the above-mentioned components separately. Finally, we summarize the findings of this section in Tables 2 and 3.

4.1. IoT and Energy Generation

Automating industrial processes and supervisory control and data acquisition systems became popular in the power sector in 1990s [37]. By monitoring and controlling equipment and processes, early stages of IoT started to contribute to the power sector by alleviating the risk of loss of production or blackout. Reliability, efficiency, environmental impacts, and maintenance issues are the main challenges of old power plants. The age of equipment in the power sector and poor maintenance problems can lead to high level of energy losses and unreliability. Assets are sometimes more than 40 years old, very expensive, and cannot be replaced easily. IoT can contribute to reducing some of these challenges in the management of power plants [37]. By applying IoT sensors, Internet-connected devices are able to distinguish any failure in the operation or abnormal decrease in energy efficiency, alarming the need for maintenance. This increases reliability and efficiency of the system, in addition to reducing the cost of maintenance [114]. According to [115], a new IoT-based power plant can save

230 million USD during the lifetime and an existing plant with the same size can save 50 million USD if equipped with the IoT platform.

Table 2. Applications of IoT in the energy sector (1): regulation, market, and energy supply side.

	Application	Sector	Description	Benefits
Regulation and market	Energy democratization	Regulation	Providing access to the grid for many small end users for peer to peer electricity trade and choosing the supplier freely.	Alleviating the hierarchy in the energy supply chain, market power, and centralized supply; liquifying the energy market and reducing the prices for consumers; and creating awareness on energy use and efficiency.
	Aggregation of small prosumers (virtual power plants)	Energy market	Aggregating load and generation of a group of end users to offer to electricity, balancing, or reserve markets.	Mobilizing small loads to participate in competitive markets; helping the grid by reducing load in peak times; Hedging the risk of high electricity bills at peak hours; and improving flexibility of the grid and reducing the need for balancing assets; Offering profitability to consumers.
Energy supply	Preventive maintenance	Upstream oil and gas industry/ utility companies	Fault, leakage, and fatigue monitoring by analyzing of big data collected through static and mobile sensors or cameras.	Reducing the risk of failure, production loss and maintenance downtime; reducing the cost of O&M; and preventing accidents and increasing safety.
	Fault maintenance	Upstream oil and gas industry/ utility companies	Identifying failures and problems in energy networks and possibly fixing them virtually.	Improving reliability of a service; improving speed in fixing leakage in district heating or failures in electricity grids; and reducing maintenance time and risk of health/safety.
	Energy storage and analytics	Industrial suppliers or utility companies	Analyzing market data and possibilities for activating flexibility options such as energy storage in the system.	Reducing the risk of supply and demand imbalance; increasing profitability in energy trade by optimal use of flexible and storage options; and ensuring an optimal strategy for storage assets.
	Digitalized power generation	Utility companies & system operator	Analyzing big data of and controlling many generation units at different time scales.	Improving security of supply; improving asset usage and management; reducing the cost of provision of backup capacity; accelerating the response to the loss of load; and reducing the risk of blackout.

For reducing fossil fuel use and relying on local energy resources, many countries are promoting RESs. Weather-dependent or variable renewable energy (VRE) sources, such as wind and solar energy, pose new challenges to the energy system known as "the intermittency challenge". In an energy system with a high share of VRE, matching generation of energy with demand is a big challenge due to variability of supply and demand resulting in mismatch in different time scales. IoT systems offer the flexibility in balancing generation with demand, which in turn can reduce the challenges of deploying VRE, resulting in higher integration shares of clean energy and less GHG emissions [116]. In addition, by employing IoT, a more efficient use of energy can be achieved by using machine-learning algorithms that help determine an optimal balance of different supply and demand technologies [37]. For instance, the use of artificial intelligence algorithms can balance the power output of a thermal power plant with the sources of in-house power generation, e.g., aggregating many small-scale solar PV panels [117].

Table 2 summarizes the applications of IoT in the energy sector, from energy supply regulation and markets.

Table 3. Applications of IoT in the energy sector (2): energy grids and demand side.

	Application	Sector	Description	Benefits
Transmission and distribution (T&D) grid	Smart grids	Electric grid management	A platform for operating the grid using big data and ICT technologies as opposed to traditional grids.	Improving energy efficiency and integration of distributed generation and load; improving security of supply; and reducing the need for backup supply capacity and costs.
	Network management	Electric grid operation & management	Using big data at different points of the grid to manage the grid more optimally.	Identifying weak points and reinforcing the grid accordingly and reducing the risk of blackout.
	Integrated control of electric vehicle fleet (EV)	Electric grid operation & management	Analyzing data of charging stations and charge/discharge cycles of EVs.	Improving the response to charging demand at peak times; analyzing and forecasting the impact of EVs on load; and identifying areas for installing new charging stations and reinforcement of the distribution grid.
	Control and management of vehicle to grid (V2G)	Electric grid operation & management	Analyzing load and charge/discharge pattern of EVs to for supporting the grid when needed.	Improving the flexibility of the system by activating EVs in supplying the grid with electricity; Reducing the need for backup capacity during peak hours Control and management of EV fleet to offer optimal interaction between the grid and EVS.
	Microgrids	Electricity grid	Platforms for managing a grid independent from the central grid.	Improving security of supply; creating interoperability and flexibility between microgrids and the main grid; and offering stable electricity prices for the consumers connected to the microgrid.
	Control and management of the District heating (DH) network	DH network	Analyzing big data of the temperature and load in the network and connected consumers.	Improving the efficiency of the grid in meeting demand; reducing the temperature of hot water supply and saving energy when possible; and identifying grid points with the need for reinforcement.
Demand side	Demand response	Residential/ commercial & industry	Central control (i.e., by shedding, shifting, or leveling.	Reducing demand at peak time, which itself reduces the grid congestion.
	Demand response (demand side management)	Residential/ commercial & industry	Central control (i.e., by shedding, shifting, or leveling; load of many consumers by analyzing the load and operation of appliances.	Reducing demand at peak time, which itself reduces the grid congestion; reducing consumer electricity bills; and reducing the need for investment in grid backup capacity.
	Advanced metering infrastructure	End users	Using sensors and devices to collect and analyze the load and temperature data in a consumer site.	Having access to detailed load variations in different time scale; identifying areas for improving energy efficiency (for example overly air-conditioned rooms or extra lights when there is no occupants); and reducing the cost of energy use.
	Battery energy management	End users	Data analytics for activating battery at the most suitable time	Optimal strategy for charge/discharge of battery in different time scale; improving energy efficiency and helping the grid at peak times; and reducing the cost of energy use.
	Smart buildings	End users	Centralized and remote control of appliances and devices.	Improving comfort by optimal control of appliances and HVAC systems; reducing manual intervention, saving time and energy; increasing knowledge on energy use and environmental impact; improving readiness for joining a smart grid or virtual power plant; and improved integration of distributed generation and storage systems.

4.2. Smart Cities

Nowadays, the staggering rate of urbanization as well as overpopulation has brought many global concerns, such as air and water pollution [118], energy access, and environmental concerns. In this line, one of the main challenges is to provide the cities with clean, affordable and reliable energy sources. The recent developments in digital technologies have provided a driving force to apply smart, IoT based solutions for the existing problems in a smart city context [119]. Smart factories, smart homes, power plants, and farms in a city can be connected and the data about their energy consumption in different hours of the day can be gathered. If it is found that a section, e.g., residential areas, consumes the most energy in the afternoon, then automatically energy devoted to other sections, e.g., factories, can be minimized to balance the whole system at a minimum cost and risk of congestion or blackout.

In a smart city, different processes, i.e., information transmission and communication, intelligent identification, location determination, tracing, monitoring, pollution control, and identity management can be managed perfectly by the aid of IoT technology [120]. IoT technologies can help to monitor every object in a city. Buildings, urban infrastructure, transport, energy networks, and utilities could be connected to sensors. These connections can ensure an energy-efficient smart city by constant monitoring of data gathered from sensors. For example, by monitoring vehicles with IoT, street lights can be controlled for optimal use of energy. In addition, the authorities can have access to the gathered information and can make more informed decisions on transportation choices and their energy demand.

4.3. Smart Grid

Smart grids are modern grids deploying the most secure and dependable ICT technology to control and optimize energy generation, T&D grids, and end usage. By connecting many smart meters, a smart grid develops a multi-directional flow of information, which can be used for optimal management of the system and efficient energy distribution [121]. The application of smart grid can be highlighted in different subsectors of the energy system individually, e.g., energy generation, buildings, or transportation, or they can be considered altogether.

In traditional grids, batteries were recharged by adapters through electricity cables and AC/DC inverter [121]. These batteries can be charged wirelessly in a smart grid, using an inductive charging technology. In addition, in a smart grid, the energy demand pattern of end users can be analyzed by collecting data through an IoT platform, for example, the time of charging of mobile phones or electric cars. Then, the nearest wireless battery charge station can allocate the right time-slot and that device/vehicle can be charged. Another advantage is that the use of IoT will lead to better control and monitoring of the battery equipped devices, and therefore, first, the energy distribution can be adjusted, and second, the delivery of electricity to these vehicles can be guaranteed. This will reduce unnecessary energy consumption considerably.

Moreover, IoT can be applied in isolated and microgrids for some islands or organizations, especially when energy is required every single moment with no exception, e.g., in databases. In such systems, all the assets connected to the grid can interact with each other. Also, the data on energy demand of any asset is accessible. This interaction can assure the perfect management of the energy distribution whenever and everywhere needed. In terms of collaborative impact of smart grids, as it is shown in Figure 3, in a smart city equipped with IoT-based smart grids, different sections of the city can be connected together [121].

During the collaborative communication between different sectors, the smart grid can alert operators through smart appliances before any acute problem occurs [113,122]. For example, through constant monitoring, it can be detected if energy demand exceeds the capacity of the grid. Therefore, by acquiring real-time data, different strategies can be adopted by authorities and energy consumption can be rescheduled to a different time when there is lower expected demand. In some regions, smart (or dynamic) pricing tariffs have been considered for variable energy prices in this regard [123]. Real-time pricing (RTP) tariffs as well as the energy price will be higher at a certain time when the consumption of energy is likely to be higher. Through the data gathered from the components of the smart grid, energy consumption and generation can be perfectly optimized and managed by far-sighted strategies. Reduction of transmission losses in T&D networks through active voltage management or reduction of non-technical losses using a network of smart meters are other examples of applying IoT [37].

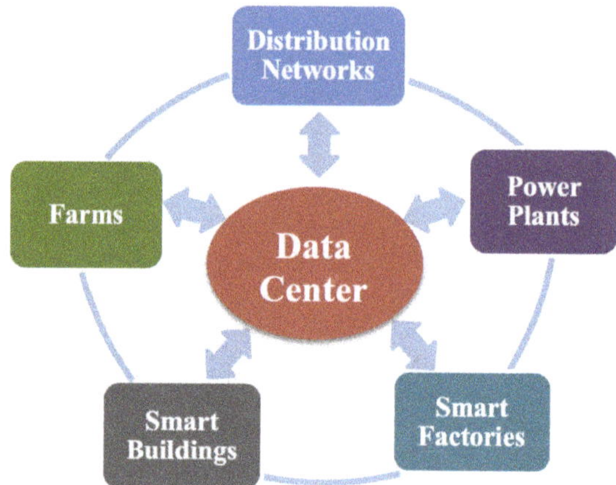

Figure 3. A centralized data connectivity in a smart city concept.

4.4. Smart Buildings

The energy consumption in cities can be divided into different parts; residential buildings (domestic); and commercial (services), including shops, offices, and schools, and transport. The domestic energy consumption in the residential sector includes lighting, equipment (appliances), domestic hot water, cooking, refrigerating, heating, ventilation, and air conditioning (HVAC) (Figure 4). HVAC energy consumption typically accounts for half of energy consumption in buildings [124]. Therefore, the management of HVAC systems is important in reducing electricity consumption. With the advancement of technology in the industry, IoT devices can play an important role to control the energy losses in HVAC systems. For example, by locating some wireless thermostats based on occupancy, unoccupied places can be realized. Once an unoccupied zone is detected, some actions can be taken to lower energy consumption. For instance, HVAC systems can reduce the operation in the unoccupied zone, which will lead to significant reduction in energy consumption and losses.

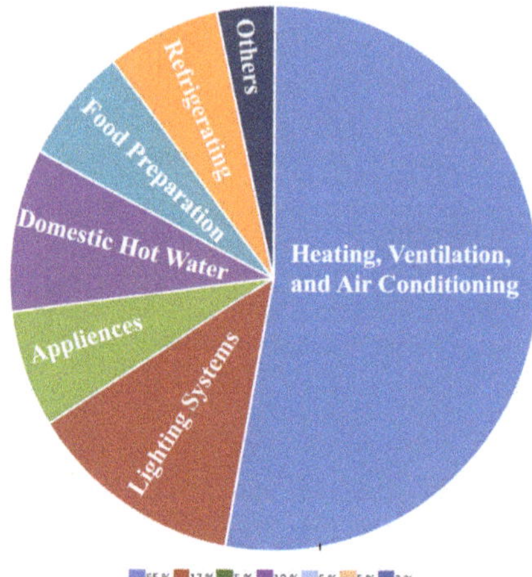

Figure 4. Share of residential energy consumption.

IoT can also be applied to manage the energy losses of lighting systems. For example, through applying IoT-based lighting systems, the customers will be alerted when the energy consumption goes beyond the standard level. Furthermore, by an efficient analysis of the real-time data, load from high-peak will be shifted to low-peak levels. This makes a significant contribution to optimal use of electrical energy [119,125] and reducing related greenhouse gas emissions. Using IoT, the demand response will be more agile and flexible, and the monitoring and demand side management will become more efficient.

4.5. Smart Use of Energy in Industry

IoT can be employed to design a fully connected and flexible system in the industry to reduce energy consumption while optimizing production. In traditional factories, a lot of energy is spent to produce the end product and control the quality of the end product. Moreover, monitoring every single process requires human resources to be involved. However, using an agile and flexible system in smart factories helps to recognize failures at the same time rather than recognizing them by monitoring the products at the end of production line. Therefore, a suitable action can be taken promptly to avert wasteful production and associated waste energy.

In terms of monitoring processes during manufacturing, IoT, and its enabling technology play a crucial role. Gateway devices, IoT hub networks, web servers, and cloud platforms, which are accessible with smart mobile devices (e.g., smart phones or personal computers) can be examples of monitoring equipment. Wireless communications such as Wi-Fi, Bluetooth, ZigBee, Z-wave, or wired communications, such as Local Area Network (LAN) can be used to connect all pieces of equipment [126]. Moreover, to use IoT more efficiently, by installing sensors on each component of an industrial site, the components that consume more energy than their nominal energy level can be detected. Thus, every single component can be easily managed, the faults of components can be fixed, and the energy consumption of each component can be optimized. This eminently results in reducing the energy losses in smart factories.

In a smart factory, data processing is the key element in the whole system, through which data in the cloud platform (acts as a brain) will be analyzed to help managers making more efficient decisions in time [127]. In terms of monitoring and maintaining assets of manufacturing, the big problem in factories is the depreciation of machines and mechanical devices. With an appropriate IoT platform and tools, the proper device size can be selected to reduce wear and tear and the associated maintenance costs. IoT-based conditional monitoring ensures the mechanical device never reaches its threshold limit. This simply means the device lasts longer and suffers fewer failures. Moreover, the failures that cause energy loss can be anticipated to be tackled.

IoT-based agile systems can provide a smart system for collaboration between customers, manufactures, and companies. Therefore, a specific product will be manufactured directly according to customers' order. Therefore, energy consumed during the process of storing spare parts as well as the energy wasted in warehouses to keep the spare parts will be dwindled significantly. Only a certain number of products in various kinds will be manufactured and stored, which enhances the management of energy consumption and production efficiency [126].

4.6. Intelligent Transportation

One of the major causes of air pollution and energy losses in big cities is overuse of private vehicles instead of public transportation. As opposed to a traditional transportation system where each system works independently, applying IoT technologies in transportation, so called "smart transportation", offers a global management system. Also, the real-time data processing plays a significant role in traffic management. All the components of the transportation system can be connected together, and their data can be processed together. Congestion control and smart parking systems using online maps are some applications of smart transportation. Smart use of transportation enables passengers to select a more cost-saving option with shorter distance and the fastest route, which saves a significant amount of time and energy [120]. Citizens will be able to determine their arrival time and manage their schedule more efficiently [125]. Therefore, time of city trips will be shortened, and the energy losses will be reduced significantly. This can remarkably reduce CO_2 emissions and other air polluting gases from transportation [119].

Table 3 summarizes the applications of IoT in the energy sector, from smart energy grids to the end use of energy. The IoT-based digitalization transforms an energy system from a unidirectional direction, i.e., from generation through energy grids to consumers, to an integrated energy system. Different parts of such an integrated smart energy system are depicted in Figure 5.

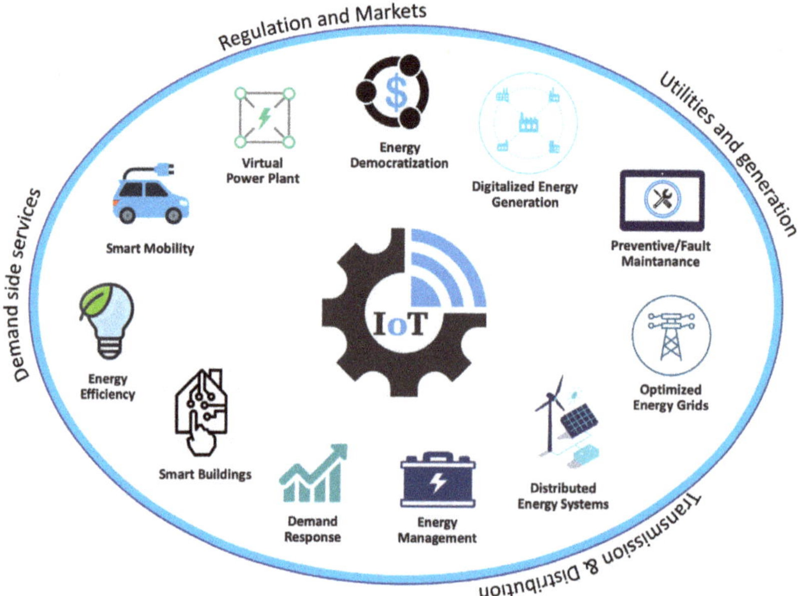

Figure 5. Applications of IoT in an integrated smart energy system.

5. Challenges of Applying IoT

Besides all the benefits of IoT for energy saving, deploying IoT in the energy sector represents challenges that need to be addressed. This section addresses the challenges and existing solutions for applying IoT-based energy systems. In addition, in Table 4, we summarize the challenges and current solutions of using IoT in the energy sector.

5.1. Energy Consumption

In the energy systems, the major effort of IoT platforms are saving the energy. In energy systems to enable communication using IoT, massive number of IoT devices transmit data. To run the IoT system and transmit huge amount of data generated from the IoT devices considerable amount of energy is needed [128]. Therefore, the energy consumption of IoT systems remains as an important challenge. However, various approaches have tried to reduce the power consumption of IoT systems. For example, by setting the sensors to go to sleep mode and just work when necessary. Designing efficient communication protocols which allow distributed computing techniques that enables energy efficient communications has been studied greatly. Applying radio optimization techniques such as modulation optimization and cooperative communication has been considered as a solution. Moreover, energy-efficient routing techniques such as cluster architectures and employing multi-path routing techniques was understood as another solution [129–131].

Table 4. Challenges and current solutions of using IoT in the energy sector.

Challenge	Issue	Example Solution	Benefit
Architecture design	Providing a reliable end-to-end connection	Using heterogeneous reference architectures	Interconnecting things and people
	Diverse technologies	Applying open standard	Scalability
Integration of IoT with subsystems	IoT data management	Designing co-simulation models	Real-time data among devices and subsystems
	Merging IoT with existing systems	Modelling integrated energy systems	Reduction in cost of maintenance
Standardization	Massive deployment of IoT devices	Defining a system of systems	Consistency among various IoT devices
	Inconsistency among IoT devices	Open information models and protocols	Covering various technologies
Energy consumption	Transmission of high data rate	Designing efficient communication protocols	Saving energy
	Efficient energy consumption	distributed computing techniques	Saving energy
IoT Security	Threats and cyber-attacks	Encryption schemes, distributed control systems	Improved security
User privacy	Maintaining users' personal information	Asking for users' permission	Enables better decision-making

5.2. Integration of IoT with Subsystems

A main challenge includes the integration of an IoT system in subsystems of the energy system. Because subsystems of the energy sector are unique employing various sensor and data communication technologies. Therefore, solutions are needed for managing the data exchange among subsystems of a IoT-enabled energy system [132–134]. An approach for finding solutions for the integration challenge, taking into account the IoT requirements of a subsystem, pertains to modeling an integrated framework for the energy system [132]. Other solutions propose designing co-simulation models for energy systems to integrate the system and minimize synchronization delay error between the subsystems [135,136].

5.3. User Privacy

Privacy refers to the right of individual or cooperative energy consumers to maintain confidentiality of their personal information when it is shared with an organization [137,138]. Therefore, accessing to proper data such as the number of energy users as well as the number and types of appliances which use energy become impossible. Indeed, these types of data which can be gathered using IoT enables better decision-making that can influence the energy production, distribution and consumption [139]. However, to decrease the violation of users' privacy, it is recommended that the energy providers ask for user permission to use their information [140], guaranteeing that the users' information will not be shared with other parties. Another solution would also be a trusted privacy management system where energy consumers have control over their information and privacy is suggested [141].

5.4. Security Challenge

The use of IoT and integration of communication technologies in energy systems enhances the threat and cyber-attacks to information of users and the energy systems from production, transmission, and distribution to consumption [142,143]. These threads define the security challenge in the energy sector [144]. Moreover, IoT-based energy systems are widely deployed in large geographical areas within the energy sector to offer services. The large deployment of IoT systems puts them in more risk of being under cyberattacks. To overcome the challenge, a study introduces an encryption scheme to secure energy information from the cyberattacks [145]. In addition, distributed control systems which enable control at different IoT system level are suggested to reduce the risk of cyberattack and increasing the security of system [146].

5.5. IoT Standards

IoT uses a variety of technologies with different standards to connect from a single device to a large number of devices. The inconsistency among IoT devices that utilize different standards forms a new challenge [147]. In IoT-enabled systems, there are two types of standards, including network protocols, communication protocols, and data-aggregation standards as well as regulatory standards related to security and privacy of data. The challenges facing the adoption of standards within IoT include the standards for handling unstructured data, security and privacy issues in addition to regulatory standards for data markets [148]. An approach for overcoming the challenge of standardization of IoT-based energy system is to define a system of systems with a common sense of understanding to allow all actors to equally access and use. Another solution pertains to developing open information models and protocols of the standards by the cooperating parties. This shall result in standards which are freely and publicly available [149].

5.6. Architecture Design

IoT-enabled systems are composed of variety of technologies with increasing number of smart interconnected devices and sensors. IoT is expected to enable communications at anytime, anywhere for any related services, generally, in an autonomic and ad hoc fashion. This means that the IoT systems based on their application purposes are designed by complex, decentralized, and mobile characteristics [149]. Taking into account the characteristics and needs of an IoT application, a reference architecture cannot be a unique solution for all of these applications. Therefore, for IoT systems, heterogeneous reference architectures are needed which are open and follow standards. The architectures also should not limit the users to use fixed and end-to-end IoT communications [149,150].

6. Future Trends

Applying current IoT systems for providing energy efficient solutions in the energy sector has many advantages highlighted in previous sections. However, for deploying IoT in the energy domain, new solutions and trends are needed to improve the performance of IoT and overcome the associated challenges. In this section, we present the Blockchain technology and Green-IoT as two approaches that can help to tackle some of the challenges.

6.1. Blockchain and IoT

Current IoT systems mostly rely on centralized cloud systems [151,152]. In most IoT applications, thousands of IoT devices and machines need to be connected, which is hard to synchronize. Moreover, due to the centralized and server-client nature of IoT when server is vulnerable, all the connected objects are easy to be hacked and compromised, which result in security concerns for the system and privacy issues for users [153]. Fortunately, Blockchain can be a solution for this challenge [154].

Blockchain provides a decentralized and democratized platform with no need for third party's intervention. The consensus platform of blockchain requires every IoT node proves that it pursues the same goal as others. Verified transactions is also stored in the form of a block, which is linked to the previous one in a way information can never be erased. Moreover, the history of every single transaction at every node can be recorded and is accessible by everyone. Therefore, any member in blockchain becomes aware of any changes in each block immediately [155–157]. Moreover, due to the distributed ledger of blockchain, even thousands of IoT devices can be synchronized easily. The consensus algorithms of blockchain based on peer-to-peer networks can provide a secure distributed database [153]. Therefore, decentralized and private-by-design IoT that can guarantee the privacy can be promised by blockchain [158].

More importantly, blockchain can store and share software updates between objects. There are innocuousness checking nodes that approve the accuracy of update information as a new node and guarantee its protection from any threats, once an update added to the blockchain as a valid block, it is impossible to erase or change it. Therefore, IoT-based platforms can be provided with updates availability and innocuousness through blockchain [159].

In the energy sector, the application of blockchain will accelerate the IoT effectiveness by providing a decentralized platform for distributed power generation and storage systems enhancing energy security and efficiency. Real and high-qualified data can be exchanged freely between devices and people can directly have access to energy information without the involvement of any third party. Neighbors can simply trade energy with one another. Therefore, without involvement of authorities, not only trust will be enhanced among people, but also many costs of this connection to the centralized grids can be saved. Another advantage is that by monitoring the usage statistics of an area, Blockchain enables the energy distribution to remotely control energy flow to that particular area. Furthermore, blockchain-based IoT systems helps in the diagnosis and maintenance of equipment within smart grid [154].

Currently, the direct application of blockchain technology in an IoT-based system is impossible due to lack of enough computational resources, insufficient bandwidth and the need to preserve power. However, cloud and fog computing platforms can ease the way for blockchain services in IoT [160].

6.2. Green IoT

The energy consumption of IoT devices is an important challenge, especially in large-scale deployment of these technologies in near future. To run billions of devices that will be connected to the Internet significant amount of energy is required. The big number of IoT devices will also produce a great deal of electronic waste [161]. To tackle these challenges, a low-carbon and efficient communication networks are needed. Fortunately, these necessities has led to the appearance of the green IoT (G-IoT) [162,163]. The key component of G-IoT is its energy-efficient characteristics throughout the life cycle, i.e., design, production, deployment, and ultimately disposal [129].

G-IoT cycle can be applied in different IoT technologies. For example, in radio frequency identification (RFID) tags. To decrease the amount of material in each RFID tag, which is difficult to be recycled, the size of RFID tags are reduced [161,164–168]. Green M2M communications is another example, which enables adjusting power transmission the minimum level, facilitates more efficient communication protocols using algorithmic and distributed computing techniques [129]. In wireless sensor networks also the sensors nodes can be in the sleep mode and just work when necessary. In addition, radio optimization techniques, such as, modulation optimization or cooperative communication can be applied to reduce the power consumption of the nodes. Moreover, energy-efficient routing techniques, such as, cluster architectures or multi-path routing can provide efficient solutions [130,131]. In conclusion, the above-mentioned approaches and examples can reduce the energy needs of IoT systems.

7. Conclusions

Energy systems are on the threshold of a new transition era. Large-scale deployment of VRE in distributed energy systems and the need for efficient use of energy calls for system-wide, integrated approaches to minimize the socio-economic-environmental impacts of energy systems. In this respect, modern technologies such as IoT can help the energy sector transform from a central, hierarchical supply chain to a decentralized, smart, and optimized system. In this paper, we review the role of IoT in the energy sector in general, and in the context of smart grids particularly.

We classify different use cases of IoT in each section of the energy supply chain, from generation through energy grids to end use sectors. The advantages of IoT-based energy management systems in increasing energy efficiency and integrating renewable energy are discussed and the findings are summarized. We discuss different components of an IoT system, including enabling communication and sensor technologies with respect to their application in the energy sector, for example, sensors of temperature, humidity, light, speed, passive infrared, and proximity. We discuss cloud computing and data analytic platforms, which are data analysis and visualization tools that can be employed for different smart applications in the energy sector, from buildings to smart cities.

We review the application of IoT in the energy supply chain under different levels, including smart cities, smart grids, smart buildings, and intelligent transportation. We discuss some of the challenges of applying IoT in the energy sector, including challenge of identifying objects, big data management, connectivity issues and uncertainty, integration of subsystems, security and privacy, energy requirements of IoT systems, standardization, and architectural design. We highlight some solutions for these challenges, i.e., Blockchain and green IoT as future directions of research.

Author Contributions: N.H.M. was involved in conceptualization, writing the paper, and designing the figures and the layout of the paper. N.H.M. supervised the content and structure of the paper and contributed to collection of data, editing and revisions. M.M was involved in writing the initial draft of the paper, and designing the figures and the layout of the paper. M.M. contributed to data collection and revisions. J.H. reviewed the paper, commented on the content and contributed to revisions. B.Z. was involved in conceptualization, reviewing the paper and commenting on the content and its relevance to the energy sector. He also contributed to supervision, data collection, editing and revisions. All authors have read and agreed to the published version of the manuscript.

Funding: This research received no external funding.

Acknowledgments: The work is supported by Helsinki Center for Data Science (HiDATA) program within Helsinki Institute for Information Technology (HIIT). The contribution of BZ was partly supported by the RE-INVEST project "Renewable Energy Investment Strategies—A two-dimensional inter-connectivity approach" funded by Innovation Fund, Denmark, and the International Institute for Applied Systems Analysis, Austria.

Conflicts of Interest: The authors declare no conflicts of interest.

References

1. Stearns, P.N. Reconceptualizing the Industrial Revolution. *J. Interdiscip. Hist.* **2011**, *42*, 442–443. [CrossRef]
2. Mokyr, J. The second industrial revolution, 1870–1914. In *Storia dell'Economia Mondiale*; Citeseer; 1998; pp. 219–245. Available online: http://citeseerx.ist.psu.edu/viewdoc/download?doi=10.1.1.481.2996&rep=rep1&type=pdf (accessed on 16 January 2020).
3. Jensen, M. The Modern Industrial Revolution, Exit, and the Failure of Internal Control Systems. *J. Financ.* **1993**, *48*, 831–880. [CrossRef]
4. Kagermann, H.; Helbig, J.; Hellinger, A.; Wahlster, W. *Recommendations for Implementing the Strategic Initiative Industrie 4.0: Securing the Future of German Manufacturing Industry*; Final Report of the Industrie 4.0 Working Group; Forschungsunion: Frankfurt/Main, Germany, 2013.
5. Witchalls, C.; Chambers, J. *The Internet of Things Business Index: A Quiet Revolution Gathers Pace*; The Economist Intelligence Unit: London, UK, 2013; pp. 58–66.
6. Datta, S.K.; Bonnet, C. MEC and IoT Based Automatic Agent Reconfiguration in Industry 4.0. In Proceedings of the 2018 IEEE International Conference on Advanced Networks and Telecommunications Systems (ANTS), Indore, India, 16–19 December 2018; pp. 1–5.
7. Shrouf, F.; Ordieres, J.; Miragliotta, G. Smart factories in Industry 4.0: A review of the concept and of energy management approached in production based on the Internet of Things paradigm. In Proceedings of the 2014 IEEE International Conference on Industrial Engineering and Engineering Management (IEEM), Selangor Darul Ehsan, Malaysia, 9–12 December 2014; pp. 697–701.
8. Bandyopadhyay, D.; Sen, J. Internet of Things: Applications and Challenges in Technology and Standardization. *Wirel. Pers. Commun.* **2011**, *58*, 49–69. [CrossRef]
9. International Energy Agency (IEA). Global Energy & CO_2 Status Report. 2019. Available online: https://www.iea.org/geco/ (accessed on 27 September 2019).

10. Intergovernmental Panel for Climate Change (IPCC). Global Warning of 1.5 °C: Summary for Policymakers. 2018. Available online: https://www.ipcc.ch/sr15/chapter/spm/ (accessed on 27 September 2019).

11. Zakeri, B.; Syri, S.; Rinne, S. Higher renewable energy integration into the existing energy system of Finland–Is there any maximum limit? *Energy* **2015**, *92*, 244–259. [CrossRef]

12. Connolly, D.; Lund, H.; Mathiesen, B. Smart Energy Europe: The technical and economic impact of one potential 100% renewable energy scenario for the European Union. *Renew. Sustain. Energy Rev.* **2016**, *60*, 1634–1653. [CrossRef]

13. Grubler, A.; Wilson, C.; Bento, N.; Boza-Kiss, B.; Krey, V.; McCollum, D.L.; Rao, N.D.; Riahi, K.; Rogelj, J.; De Stercke, S.; et al. A low energy demand scenario for meeting the 1.5 C target and sustainable development goals without negative emission technologies. *Nat. Energy* **2018**, *3*, 515–527.

14. UN. *Special Edition: Progress towards the Sustainable Development Goals*; UN: New York, NY, USA, 2019.

15. Tan, Y.S.; Ng, Y.T.; Low, J.S.C. Internet-of-things enabled real-time monitoring of energy efficiency on manufacturing shop floors. *Procedia CIRP* **2017**, *61*, 376–381. [CrossRef]

16. Bhattacharyya, S.C. *Energy Economics: Concepts, Issues, Markets and Governance*; Springer: Berlin/Heidelberg, Germany, 2011.

17. Tamilselvan, K.; Thangaraj, P. Pods—A novel intelligent energy efficient and dynamic frequency scalings for multi-core embedded architectures in an IoT environment. *Microprocess. Microsyst.* **2020**, *72*, 102907. [CrossRef]

18. Zhou, K.; Yang, S.; Shao, Z. Energy Internet: The business perspective. *Appl. Energy* **2016**, *178*, 212–222. [CrossRef]

19. Motlagh, N.H.; Khajavi, S.H.; Jaribion, A.; Holmstrom, J. An IoT-based automation system for older homes: A use case for lighting system. In Proceedings of the 2018 IEEE 11th Conference on Service-Oriented Computing and Applications (SOCA), Paris, France, 19–22 November 2018; pp. 1–6.

20. Da Xu, L.; He, W.; Li, S. Internet of Things in Industries: A Survey. *IEEE Trans. Ind. Inform.* **2014**, *10*, 2233–2243.

21. Talari, S.; Shafie-Khah, M.; Siano, P.; Loia, V.; Tommasetti, A.; Catalão, J. A review of smart cities based on the internet of things concept. *Energies* **2017**, *10*, 421. [CrossRef]

22. Ibarra-Esquer, J.; González-Navarro, F.; Flores-Rios, B.; Burtseva, L.; Astorga-Vargas, M. Tracking the evolution of the internet of things concept across different application domains. *Sensors* **2017**, *17*, 1379. [CrossRef]

23. Swan, M. Sensor mania! the internet of things, wearable computing, objective metrics, and the quantified self 2.0. *J. Sens. Actuator Netw.* **2012**, *1*, 217–253. [CrossRef]

24. Gupta, A.; Jha, R.K. A survey of 5G network: Architecture and emerging technologies. *IEEE Access* **2015**, *3*, 1206–1232. [CrossRef]

25. Stojkoska, B.L.R.; Trivodaliev, K.V. A review of Internet of Things for smart home: Challenges and solutions. *J. Clean. Prod.* **2017**, *140*, 1454–1464. [CrossRef]

26. Hui, H.; Ding, Y.; Shi, Q.; Li, F.; Song, Y.; Yan, J. 5G network-based Internet of Things for demand response in smart grid: A survey on application potential. *Appl. Energy* **2020**, *257*, 113972. [CrossRef]

27. Petroşanu, D.M.; Căruţaşu, G.; Căruţaşu, N.L.; Pîrjan, A. A Review of the Recent Developments in Integrating Machine Learning Models with Sensor Devices in the Smart Buildings Sector with a View to Attaining Enhanced Sensing, Energy Efficiency, and Optimal Building Management. *Energies* **2019**, *12*, 4745. [CrossRef]

28. Luo, X.G.; Zhang, H.B.; Zhang, Z.L.; Yu, Y.; Li, K. A New Framework of Intelligent Public Transportation System Based on the Internet of Things. *IEEE Access* **2019**, *7*, 55290–55304. [CrossRef]

29. Khatua, P.K.; Ramachandaramurthy, V.K.; Kasinathan, P.; Yong, J.Y.; Pasupuleti, J.; Rajagopalan, A. Application and Assessment of Internet of Things toward the Sustainability of Energy Systems: Challenges and Issues. *Sustain. Cities Soc.* **2019**, 101957. [CrossRef]

30. Haseeb, K.; Almogren, A.; Islam, N.; Ud Din, I.; Jan, Z. An Energy-Efficient and Secure Routing Protocol for Intrusion Avoidance in IoT-Based WSN. *Energies* **2019**, *12*, 4174. [CrossRef]

31. Zouinkhi, A.; Ayadi, H.; Val, T.; Boussaid, B.; Abdelkrim, M.N. Auto-management of energy in IoT networks. *Int. J. Commun. Syst.* **2019**, *33*, e4168. [CrossRef]

32. Höller, J.; Tsiatsis, V.; Mulligan, C.; Avesand, S.; Karnouskos, S.; Boyle, D. *From Machine-to-Machine to the Internet of Things: Introduction to a New Age of Intelligence*; Elsevier: Amsterdam, The Netherlands, 2014.

33. Atzori, L.; Iera, A.; Morabito, G. The Internet of Things: A survey. *Comput. Netw.* **2010**, *54*, 2787–2805. [CrossRef]

34. Hui, T.K.; Sherratt, R.S.; Sánchez, D.D. Major requirements for building Smart Homes in Smart Cities based on Internet of Things technologies. *Future Gener. Comput. Syst.* **2017**, *76*, 358–369. [CrossRef]

35. Evans, D. The Internet of Things: How the Next Evolution of the Internet is Changing Everything. *CISCO White Pap.* **2011**, *1*, 1–11.

36. Motlagh, N.H.; Bagaa, M.; Taleb, T. Energy and Delay Aware Task Assignment Mechanism for UAV-Based IoT Platform. *IEEE Internet Things J.* **2019**, *6*, 6523–6536. [CrossRef]

37. Ramamurthy, A.; Jain, P. *The Internet of Things in the Power Sector: Opportunities in Asia and the Pacific*; Asian Development Bank: Mandaluyong, Philippines, 2017.

38. Jia, M.; Komeily, A.; Wang, Y.; Srinivasan, R.S. Adopting Internet of Things for the development of smart buildings: A review of enabling technologies and applications. *Autom. Constr.* **2019**, *101*, 111–126. [CrossRef]

39. Karunarathne, G.R.; Kulawansa, K.T.; Firdhous, M.M. Wireless Communication Technologies in Internet of Things: A Critical Evaluation. In Proceedings of the 2018 International Conference on Intelligent and Innovative Computing Applications (ICONIC), Plaine Magnien, Mauritius, 6–7 December 2018; pp. 1–5.

40. Li, S.; Da Xu, L.; Zhao, S. 5G Internet of Things: A survey. *J. Ind. Inf. Integr.* **2018**, *10*, 1–9. [CrossRef]

41. Watson Internet of Things. Securely Connect with Watson IoT Platform. 2019. Available online: https://www.ibm.com/internet-of-things/solutions/iot-platform/watson-iot-platform (accessed on 15 October 2019).

42. Kelly, S.D.T.; Suryadevara, N.K.; Mukhopadhyay, S.C. Towards the Implementation of IoT for Environmental Condition Monitoring in Homes. *IEEE Sens. J.* **2013**, *13*, 3846–3853. [CrossRef]

43. Newark Element. Smart Sensor Technology for the IoT. 2018. Available online: https://www.techbriefs.com/component/content/article/tb/features/articles/33212 (accessed on 25 December 2019).

44. Rault, T.; Bouabdallah, A.; Challal, Y. Energy efficiency in wireless sensor networks: A top-down survey. *Comput. Netw.* **2014**, *67*, 104–122. [CrossRef]

45. Di Francia, G. The development of sensor applications in the sectors of energy and environment in Italy, 1976–2015. *Sensors* **2017**, *17*, 793. [CrossRef]

46. ITFirms Co. 8 Types of Sensors that Coalesce Perfectly with an IoT App. 2018. Available online: https://www.itfirms.co/8-types-of-sensors-that-coalesce-perfectly-with-an-iot-app/ (accessed on 27 September 2019).

47. Morris, A.S.; Langari, R. Level Measurement. In *Measurement and Instrumentation*, 2nd ed.; Morris, A.S., Langari, R., Eds.; Academic Press: Boston, MA, USA, 2016; Chapter 17, pp. 531–545.

48. Pérez-Lombard, L.; Ortiz, J.; Pout, C. A review on buildings energy consumption information. *Energy Build.* **2008**, *40*, 394–398. [CrossRef]

49. Moram, M. Lighting Up Lives with Energy Efficient Lighting. 2012. Available online: http://aglobalvillage.org/journal/issue7/waste/lightinguplives/ (accessed on 27 December 2019).

50. Riyanto, I.; Margatama, L.; Hakim, H.; Hindarto, D. Motion Sensor Application on Building Lighting Installation for Energy Saving and Carbon Reduction Joint Crediting Mechanism. *Appl. Syst. Innov.* **2018**, *1*, 23. [CrossRef]

51. Kim, W.; Mechitov, K.; Choi, J.; Ham, S. On target tracking with binary proximity sensors. In Proceedings of the IPSN 2005—Fourth International Symposium on Information Processing in Sensor Networks, Los Angeles, CA, USA, 25–27 April 2005; pp. 301–308.

52. Pepperl+Fuchs. Sensors for Wind Energy Applications. 2019. Available online: https://www.pepperl-fuchs.com/global/en/15351.htm (accessed on 27 December 2019).

53. Kececi, E.F. Actuators. In *Mechatronic Components*; Kececi, E.F., Ed.; Butterworth-Heinemann: Oxford, UK, 2019; Chapter 11, pp. 145–154.

54. Nesbitt, B. *Handbook of Valves and Actuators: Valves Manual International*; Elsevier: Amsterdam, The Netherlands, 2011.

55. Ray, R. Valves and Actuators. *Power Eng.* **2014**, *118*, 4862.

56. Blanco, J.; García, A.; Morenas, J. Design and Implementation of a Wireless Sensor and Actuator Network to Support the Intelligent Control of Efficient Energy Usage. *Sensors* **2018**, *18*, 1892. [CrossRef] [PubMed]

57. Martínez-Cruz.; Eugenio, C. Manufacturing low-cost wifi-based electric energy meter. In Proceedings of the 2014 IEEE Central America and Panama Convention (CONCAPAN), Panama City, Panama, 12–14 November 2014; pp. 1–6.

58. Rodriguez-Diaz, E.; Vasquez, J.C.; Guerrero, J.M. Intelligent DC Homes in Future Sustainable Energy Systems: When efficiency and intelligence work together. *IEEE Consum. Electron. Mag.* **2016**, *5*, 74–80. [CrossRef]

59. Karthika, A.; Valli, K.R.; Srinidhi, R.; Vasanth, K. Automation Of Energy Meter And Building A Network Using Iot. In Proceedings of the 2019 5th International Conference on Advanced Computing Communication Systems (ICACCS), Coimbatore, India, 15–16 March 2019; pp. 339–341.

60. Lee, T.; Jeon, S.; Kang, D.; Park, L.W.; Park, S. Design and implementation of intelligent HVAC system based on IoT and Big data platform. In Proceedings of the 2017 IEEE International Conference on Consumer Electronics (ICCE), Las Vegas, NV, USA, 8–10 January 2017; pp. 398–399.

61. Lee, Y.; Hsiao, W.; Huang, C.; Chou, S.T. An integrated cloud-based smart home management system with community hierarchy. *IEEE Trans. Consum. Electron.* **2016**, *62*, 1–9. [CrossRef]

62. Kabalci, Y.; Kabalci, E.; Padmanaban, S.; Holm-Nielsen, J.B.; Blaabjerg, F. Internet of Things applications as energy internet in Smart Grids and Smart Environments. *Electronics* **2019**, *8*, 972. [CrossRef]

63. Jain, S.; Pradish, M.; Paventhan, A.; Saravanan, M.; Das, A. Smart Energy Metering Using LPWAN IoT Technology. In *ISGW 2017: Compendium of Technical Papers*; Springer: Berlin/Heidelberg, Germany, 2018; pp. 19–28.

64. Lee, J.; Su, Y.; Shen, C. A Comparative Study of Wireless Protocols: Bluetooth, UWB, ZigBee, and Wi-Fi. In Proceedings of the IECON 2007—33rd Annual Conference of the IEEE Industrial Electronics Society, Taipei, Taiwan, 5–8 November 2007; pp. 46–51.

65. Choi, M.; Park, W.; Lee, I. Smart office energy management system using bluetooth low energy based beacons and a mobile app. In Proceedings of the 2015 IEEE International Conference on Consumer Electronics (ICCE), Las Vegas, NV, USA, 9–12 January 2015; pp. 501–502.

66. Collotta, M.; Pau, G. A Novel Energy Management Approach for Smart Homes Using Bluetooth Low Energy. *IEEE J. Sel. Areas Commun.* **2015**, *33*, 2988–2996. [CrossRef]

67. Collotta, M.; Pau, G. A solution based on bluetooth low energy for smart home energy management. *Energies* **2015**, *8*, 11916–11938. [CrossRef]

68. Craig, W.C. *Zigbee: Wireless Control that Simply Works*; Zigbee Alliance ZigBee Alliance: Davis, CA, USA, 2004.

69. Froiz-Míguez, I.; Fernández-Caramés, T.; Fraga-Lamas, P.; Castedo, L. Design, implementation and practical evaluation of an IoT home automation system for fog computing applications based on MQTT and ZigBee-WiFi sensor nodes. *Sensors* **2018**, *18*, 2660. [CrossRef]

70. Erol-Kantarci, M.; Mouftah, H.T. Wireless Sensor Networks for Cost-Efficient Residential Energy Management in the Smart Grid. *IEEE Trans. Smart Grid* **2011**, *2*, 314–325. [CrossRef]

71. Han, D.; Lim, J. Smart home energy management system using IEEE 802.15.4 and zigbee. *IEEE Trans. Consum. Electron.* **2010**, *56*, 1403–1410. [CrossRef]

72. Han, J.; Choi, C.; Park, W.; Lee, I.; Kim, S. Smart home energy management system including renewable energy based on ZigBee and PLC. In Proceedings of the 2014 IEEE International Conference on Consumer Electronics (ICCE), Las Vegas, NV, USA, 4–6 January 2014; pp. 544–545.

73. Batista, N.; Melício, R.; Matias, J.; Catalão, J. Photovoltaic and wind energy systems monitoring and building/home energy management using ZigBee devices within a smart grid. *Energy* **2013**, *49*, 306–315. [CrossRef]

74. Augustin, A.; Yi, J.; Clausen, T.; Townsley, W. A study of LoRa: Long range & low power networks for the internet of things. *Sensors* **2016**, *16*, 1466.

75. Mataloto, B.; Ferreira, J.C.; Cruz, N. LoBEMS—IoT for Building and Energy Management Systems. *Electronics* **2019**, *8*, 763. [CrossRef]

76. Javed, A.; Larijani, H.; Wixted, A. Improving Energy Consumption of a Commercial Building with IoT and Machine Learning. *IT Prof.* **2018**, *20*, 30–38. [CrossRef]

77. Ferreira, J.C.; Afonso, J.A.; Monteiro, V.; Afonso, J.L. An Energy Management Platform for Public Buildings. *Electronics* **2018**, *7*, 294. [CrossRef]

78. Gomez, C.; Veras, J.C.; Vidal, R.; Casals, L.; Paradells, J. A Sigfox energy consumption model. *Sensors* **2019**, *19*, 681. [CrossRef]

79. Pitì, A.; Verticale, G.; Rottondi, C.; Capone, A.; Lo Schiavo, L. The role of smart meters in enabling real-time energy services for households: The Italian case. *Energies* **2017**, *10*, 199. [CrossRef]

80. Mekki, K.; Bajic, E.; Chaxel, F.; Meyer, F. Overview of Cellular LPWAN Technologies for IoT Deployment: Sigfox, LoRaWAN, and NB-IoT. In Proceedings of the 2018 IEEE International Conference on Pervasive Computing and Communications Workshops (PerCom Workshops), Athens, Greece, 19–23 March 2018; pp. 197–202.

81. Nair, V.; Litjens, R.; Zhang, H. Optimisation of NB-IoT deployment for smart energy distribution networks. *Eurasip J. Wirel. Commun. Netw.* **2019**, *2019*, 186. [CrossRef]

82. Pennacchioni, M.; Di Benedette, M.; Pecorella, T.; Carlini, C.; Obino, P. NB-IoT system deployment for smart metering: Evaluation of coverage and capacity performances. In Proceedings of the 2017 AEIT International Annual Conference, Cagliari, Italy, 20–22 September 2017; pp. 1–6.

83. Li, Y.; Cheng, X.; Cao, Y.; Wang, D.; Yang, L. Smart Choice for the Smart Grid: Narrowband Internet of Things (NB-IoT). *IEEE Internet Things J.* **2018**, *5*, 1505–1515. [CrossRef]

84. Shariatmadari, H.; Ratasuk, R.; Iraji, S.; Laya, A.; Taleb, T.; Jäntti, R.; Ghosh, A. Machine-type communications: Current status and future perspectives toward 5G systems. *IEEE Commun. Mag.* **2015**, *53*, 10–17. [CrossRef]

85. Lauridsen, M.; Kovacs, I.Z.; Mogensen, P.; Sorensen, M.; Holst, S. Coverage and Capacity Analysis of LTE-M and NB-IoT in a Rural Area. In Proceedings of the 2016 IEEE 84th Vehicular Technology Conference (VTC-Fall), Montreal, QC, Canada, 18–21 September 2016; pp. 1–5.

86. Deshpande, K.V.; Rajesh, A. Investigation on imcp based clustering in lte-m communication for smart metering applications. *Eng. Sci. Technol. Int. J.* **2017**, *20*, 944–955. [CrossRef]

87. Emmanuel, M.; Rayudu, R. Communication technologies for smart grid applications: A survey. *J. Netw. Comput. Appl.* **2016**, *74*, 133–148. [CrossRef]

88. Webb, W. Weightless: The technology to finally realise the M2M vision. *Int. J. Interdiscip. Telecommun. Netw. (IJITN)* **2012**, *4*, 30–37. [CrossRef]

89. Sethi, P.; Sarangi, S.R. Internet of things: Architectures, protocols, and applications. *J. Electr. Comput. Eng.* **2017**, *2017*. [CrossRef]

90. Wei, J.; Han, J.; Cao, S. Satellite IoT Edge Intelligent Computing: A Research on Architecture. *Electronics* **2019**, *8*, 1247. [CrossRef]

91. Sohraby, K.; Minoli, D.; Occhiogrosso, B.; Wang, W. A review of wireless and satellite-based m2m/iot services in support of smart grids. *Mob. Networks Appl.* **2018**, *23*, 881–895. [CrossRef]

92. De Sanctis, M.; Cianca, E.; Araniti, G.; Bisio, I.; Prasad, R. Satellite Communications Supporting Internet of Remote Things. *IEEE Internet Things J.* **2016**, *3*, 113–123. [CrossRef]

93. GSMA. *Security Features of LTE-M and NB-IoT Networks*; Technical Report; GSMA: London, UK, 2019.

94. Sigfox. *Make Things Come Alive in a Secure Way*; Technical Report; Sigfox: Labège, France, 2017.

95. Sanchez-Iborra, R.; Cano, M.D. State of the art in LP-WAN solutions for industrial IoT services. *Sensors* **2016**, *16*, 708. [CrossRef]

96. Siekkinen, M.; Hiienkari, M.; Nurminen, J.K.; Nieminen, J. How low energy is bluetooth low energy? comparative measurements with zigbee/802.15. 4. In Proceedings of the 2012 IEEE Wireless Communications and Networking Conference workshops (WCNCW), Paris, France, 1 April 2012; pp. 232–237.

97. Lee, J.S.; Dong, M.F.; Sun, Y.H. A preliminary study of low power wireless technologies: ZigBee and Bluetooth low energy. In Proceedings of the 2015 IEEE 10th Conference on Industrial Electronics and Applications (ICIEA), Auckland, New Zealand, 15–17 June 2015; pp. 135–139.

98. Fraire, J.A.; Céspedes, S.; Accettura, N. Direct-To-Satellite IoT-A Survey of the State of the Art and Future Research Perspectives. In Proceedings of the 2019 International Conference on Ad-Hoc Networks and Wireless, Luxembourg, 1–3 October 2019; pp. 241–258.

99. Jaribion, A.; Khajavi, S.H.; Hossein Motlagh, N.; Holmström, J. [WiP] A Novel Method for Big Data Analytics and Summarization Based on Fuzzy Similarity Measure. In Proceedings of the 2018 IEEE 11th Conference on Service-Oriented Computing and Applications (SOCA), Paris, France, 19–22 November 2018; pp. 221–226.

100. Chen, M.; Mao, S.; Liu, Y. Big Data: A Survey. *Mob. Netw. Appl.* **2014**, *19*, 171–209. [CrossRef]

101. Stojmenovic, I. Machine-to-Machine Communications With In-Network Data Aggregation, Processing, and Actuation for Large-Scale Cyber-Physical Systems. *IEEE Internet Things J.* **2014**, *1*, 122–128. [CrossRef]

102. Chen, H.; Chiang, R.H.; Storey, V.C. Business intelligence and analytics: From big data to big impact. *MIS Q. Manag. Inf. Syst.* **2012**, *36*, 1165–1188. [CrossRef]

103. Intel IT Centre. *Big Data Analytics: Intel's IT Manager Survey on How Organizations Are Using Big Data*; Technical Report; Intel IT Centre: Santa Clara, CA, USA, 2012.

104. Stergiou, C.; Psannis, K.E.; Kim, B.G.; Gupta, B. Secure integration of IoT and Cloud Computing. *Future Gener. Comput. Syst.* **2018**, *78*, 964–975. [CrossRef]

105. Josep, A.D.; Katz, R.; Konwinski, A.; Gunho, L.; Patterson, D.; Rabkin, A. A view of cloud computing. *Commun. ACM* **2010**, *53*. [CrossRef]

106. Ji, C.; Li, Y.; Qiu, W.; Awada, U.; Li, K. Big Data Processing in Cloud Computing Environments. In Proceedings of the 2012 12th International Symposium on Pervasive Systems, Algorithms and Networks, San Marcos, TX, USA, 13–15 December 2012; pp. 17–23.

107. Foster, I.; Zhao, Y.; Raicu, I.; Lu, S. Cloud Computing and Grid Computing 360-Degree Compared. In Proceedings of the 2008 Grid Computing Environments Workshop, Austin, TX, USA, 16 November 2008; pp. 1–10.

108. Hamdaqa, M.; Tahvildari, L. *Cloud Computing Uncovered: A Research Landscape*; Advances in Computers; Elsevier: Amsterdam, The Netherlands, 2012; Volume 86, pp. 41–85.

109. Khan, Z.; Anjum, A.; Kiani, S.L. Cloud Based Big Data Analytics for Smart Future Cities. In Proceedings of the 2013 IEEE/ACM 6th International Conference on Utility and Cloud Computing, Dresden, Germany, 9–12 December 2013; pp. 381–386.

110. Mahmud, R.; Kotagiri, R.; Buyya, R. Fog computing: A taxonomy, survey and future directions. In *Internet of Everything*; Springer: Berlin/Heidelberg, Germany, 2018; pp. 103–130.

111. Verma, M.; Bhardwaj, N.; Yadav, A.K. Real time efficient scheduling algorithm for load balancing in fog computing environment. *Int. J. Comput. Sci. Inf. Technol.* **2016**, *8*, 1–10. [CrossRef]

112. Atlam, H.F.; Walters, R.J.; Wills, G.B. Fog computing and the internet of things: A review. *Big Data Cogn. Comput.* **2018**, *2*, 10. [CrossRef]

113. Bhardwaj, A. *Leveraging the Internet of Things and Analytics for Smart Energy Management*; TATA Consultancy Services: Mumbai, India, 2015.

114. Sigfox, Inc. Utilities & Energy. 2019. Available online: https://www.sigfox.com/en/utilities-energy/ (accessed on 27 September 2019).

115. Immelt, J.R. *The Future of Electricity Is Digital*; Technical Report; General Electric: Boston, MA, USA, 2015.

116. Al-Ali, A. Internet of things role in the renewable energy resources. *Energy Procedia* **2016**, *100*, 34–38. [CrossRef]

117. Karnouskos, S. The cooperative internet of things enabled smart grid. In Proceedings of the 14th IEEE International Symposium on Consumer Electronics (ISCE2010), Braunschweig, Germany, 7–10 June 2010; pp. 7–10.

118. Lagerspetz, E.; Motlagh, N.H.; Zaidan, M.A.; Fung, P.L.; Mineraud, J.; Varjonen, S.; Siekkinen, M.; Nurmi, P.; Matsumi, Y.; Tarkoma, S.; et al. MegaSense: Feasibility of Low-Cost Sensors for Pollution Hot-spot Detection. In Proceedings of the 2019 IEEE 17th International Conference on Industrial Informatics (INDIN), Helsinki-Espoo, Finland, 23–25 July 2019.

119. Ejaz, W.; Naeem, M.; Shahid, A.; Anpalagan, A.; Jo, M. Efficient energy management for the internet of things in smart cities. *IEEE Commun. Mag.* **2017**, *55*, 84–91. [CrossRef]

120. Mohanty, S.P. Everything you wanted to know about smart cities: The Internet of things is the backbone. *IEEE Consum. Electron. Mag.* **2016**, *5*, 60–70. [CrossRef]

121. Hossain, M.; Madlool, N.; Rahim, N.; Selvaraj, J.; Pandey, A.; Khan, A.F. Role of smart grid in renewable energy: An overview. *Renew. Sustain. Energy Rev.* **2016**, *60*, 1168–1184. [CrossRef]

122. Karnouskos, S.; Colombo, A.W.; Lastra, J.L.M.; Popescu, C. Towards the energy efficient future factory. In Proceedings of the 2009 7th IEEE International Conference on Industrial Informatics, Cardiff, UK, 23–26 June 2009; pp. 367–371.

123. M. Avci, M.E.; Asfour, S. Residential HVAC load control strategy in real-time electricity pricing environment. In Proceedings of the 2012 IEEE Conference on Energytech, Cleveland, OH, USA, 29–31 May 2012; pp. 1–6.

124. Vakiloroaya, V.; Samali, B.; Fakhar, A.; Pishghadam, K. A review of different strategies for HVAC energy saving. *Energy Convers. Manag.* **2014**, *77*, 738–754. [CrossRef]

125. Arasteh, H.; Hosseinnezhad, V.; Loia, V.; Tommasetti, A.; Troisi, O.; Shafie-khah, M.; Siano, P. IoT-based smart cities: A survey. In Proceedings of the 2016 IEEE 16th International Conference on Environment and Electrical Engineering (EEEIC), Florence, Italy, 7–10 June 2016; pp. 1–6.

126. Lee, C.; Zhang, S. Development of an Industrial Internet of Things Suite for Smart Factory towards Re-industrialization in Hong Kong. In Proceedings of the 6th International Workshop of Advanced Manufacturing and Automation, Manchester, UK, 10–11 November 2016.

127. Reinfurt, L.; Falkenthal, M.; Breitenbücher, U.; Leymann, F. Applying IoT Patterns to Smart Factory Systems. In Proceedings of the 2017 Advanced Summer School on Service Oriented Computing (Summer SOC), Hersonissos, Greece, 25–30 June 2017.

128. Kaur, N.; Sood, S.K. An energy-efficient architecture for the Internet of Things (IoT). *IEEE Syst. J.* **2015**, *11*, 796–805. [CrossRef]

129. Shaikh, F.K.; Zeadally, S.; Exposito, E. Enabling technologies for green internet of things. *IEEE Syst. J.* **2015**, *11*, 983–994. [CrossRef]

130. Lin, Y.; Chou, Z.; Yu, C.; Jan, R. Optimal and Maximized Configurable Power Saving Protocols for Corona-Based Wireless Sensor Networks. *IEEE Trans. Mob. Comput.* **2015**, *14*, 2544–2559. [CrossRef]

131. Anastasi, G.; Conti, M.; Di Francesco, M.; Passarella, A. Energy conservation in wireless sensor networks: A survey. *Ad Hoc Netw.* **2009**, *7*, 537–568. [CrossRef]

132. Shakerighadi, B.; Anvari-Moghaddam, A.; Vasquez, J.C.; Guerrero, J.M. Internet of Things for Modern Energy Systems: State-of-the-Art, Challenges, and Open Issues. *Energies* **2018**, *11*, 1252. [CrossRef]

133. Anjana, K.; Shaji, R. A review on the features and technologies for energy efficiency of smart grid. *Int. J. Energy Res.* **2018**, *42*, 936–952. [CrossRef]

134. Boroojeni, K.; Amini, M.H.; Nejadpak, A.; Dragičević, T.; Iyengar, S.S.; Blaabjerg, F. A Novel Cloud-Based Platform for Implementation of Oblivious Power Routing for Clusters of Microgrids. *IEEE Access* **2017**, *5*, 607–619. [CrossRef]

135. Kounev, V.; Tipper, D.; Levesque, M.; Grainger, B.M.; Mcdermott, T.; Reed, G.F. A microgrid co-simulation framework. In Proceedings of the 2015 Workshop on Modeling and Simulation of Cyber-Physical Energy Systems (MSCPES), Seattle, WA, USA, 13 April 2015; pp. 1–6.

136. Wong, T.Y.; Shum, C.; Lau, W.H.; Chung, S.; Tsang, K.F.; Tse, C. Modeling and co-simulation of IEC61850-based microgrid protection. In Proceedings of the 2016 IEEE International Conference on Smart Grid Communications (SmartGridComm), Sydney, Australia, 6–9 November 2016; pp. 582–587.

137. Porambage, P.; Ylianttila, M.; Schmitt, C.; Kumar, P.; Gurtov, A.; Vasilakos, A.V. The quest for privacy in the internet of things. *IEEE Cloud Comput.* **2016**, *3*, 36–45. [CrossRef]

138. Chow, R. The Last Mile for IoT Privacy. *IEEE Secur. Priv.* **2017**, *15*, 73–76. [CrossRef]

139. Jayaraman, P.P.; Yang, X.; Yavari, A.; Georgakopoulos, D.; Yi, X. Privacy preserving Internet of Things: From privacy techniques to a blueprint architecture and efficient implementation. *Future Gener. Comput. Syst.* **2017**, *76*, 540–549. [CrossRef]

140. Roman, R.; Najera, P.; Lopez, J. Securing the internet of things. *Computer* **2011**, *44*, 51–58. [CrossRef]

141. Fhom, H.S.; Kuntze, N.; Rudolph, C.; Cupelli, M.; Liu, J.; Monti, A. A user-centric privacy manager for future energy systems. In Proceedings of the 2010 International Conference on Power System Technology, Hangzhou, China, 24–28 October 2010; pp. 1–7.

142. Dorri, A.; Kanhere, S.S.; Jurdak, R.; Gauravaram, P. Blockchain for IoT security and privacy: The case study of a smart home. In Proceedings of the 2017 IEEE International Conference on Pervasive Computing and Communications Workshops (PerCom Workshops), Kona, HI, USA, 13–17 March 2017; pp. 618–623.

143. Poyner, I.; Sherratt, R.S. Privacy and security of consumer IoT devices for the pervasive monitoring of vulnerable people. In Proceedings of the Living in the Internet of Things: Cybersecurity of the IoT—2018, London, UK, 28–29 March 2018; pp. 1–5.

144. Li, Z.; Shahidehpour, M.; Aminifar, F. Cybersecurity in distributed power systems. *Proc. IEEE* **2017**, *105*, 1367–1388. [CrossRef]

145. Song, T.; Li, R.; Mei, B.; Yu, J.; Xing, X.; Cheng, X. A privacy preserving communication protocol for IoT applications in smart homes. *IEEE Internet Things J.* **2017**, *4*, 1844–1852. [CrossRef]

146. Roman, R.; Lopez, J. Security in the distributed internet of things. In Proceedings of the 2012 International Conference on Trusted Systems, London, UK, 17–18 December 2012; pp. 65–66.

147. Meddeb, A. Internet of things standards: Who stands out from the crowd? *IEEE Commun. Mag.* **2016**, *54*, 40–47. [CrossRef]

148. Banafa, A. IoT Standardization and Implementation Challenges. 2016. Available online: https://iot.ieee.org/newsletter/july-2016/iot-standardization-and-implementation-challenges.html (accessed on 10 May 2019).

149. Chen, S.; Xu, H.; Liu, D.; Hu, B.; Wang, H. A Vision of IoT: Applications, Challenges, and Opportunities With China Perspective. *IEEE Internet Things J.* **2014**, *1*, 349–359. [CrossRef]

150. Al-Qaseemi, S.A.; Almulhim, H.A.; Almulhim, M.F.; Chaudhry, S.R. IoT architecture challenges and issues: Lack of standardization. In Proceedings of the 2016 Future Technologies Conference (FTC), San Francisco, CA, USA, 6–7 December 2016; pp. 731–738.

151. Kshetri, N. Can Blockchain Strengthen the Internet of Things? *IT Prof.* **2017**, *19*, 68–72. [CrossRef]

152. Dorri, A.; Kanhere, S.S.; Jurdak, R. Towards an optimized blockchain for IoT. In Proceedings of the Second International Conference on Internet-of-Things Design and Implementation, Pittsburgh, PA, USA, 18–21 April 2017; pp. 173–178.

153. Huh, S.; Cho, S.; Kim, S. Managing IoT devices using blockchain platform. In Proceedings of the 2017 19th International Conference on Advanced Communication Technology (ICACT), Bongpyeong, Korea, 19–22 February 2017; pp. 464–467.

154. Alladi, T.; Chamola, V.; Rodrigues, J.J.; Kozlov, S.A. Blockchain in Smart Grids: A Review on Different Use Cases. *Sensors* **2019**, *19*, 4862. [CrossRef]

155. Christidis, K.; Devetsikiotis, M. Blockchains and Smart Contracts for the Internet of Things. *IEEE Access* **2016**, *4*, 2292–2303. [CrossRef]

156. Korpela, K.; Hallikas, J.; Dahlberg, T. Digital Supply Chain Transformation toward Blockchain Integration. In Proceedings of the 50th Hawaii International Conference on Ssystem Sciences, Waikoloa, HI, USA, 4–7 January 2017.

157. Hawlitschek, F.; Notheisen, B.; Teubner, T. The limits of trust-free systems: A literature review on blockchain technology and trust in the sharing economy. *Electron. Commer. Res. Appl.* **2018**, *29*, 50–63. [CrossRef]

158. Conoscenti, M.; Vetro, A.; De Martin, J.C. Blockchain for the Internet of Things: A systematic literature review. In Proceedings of the 2016 IEEE/ACS 13th International Conference of Computer Systems and Applications (AICCSA), Agadir, Morocco, 29 November–2 December 2016; pp. 1–6.

159. Boudguiga, A.; Bouzerna, N.; Granboulan, L.; Olivereau, A.; Quesnel, F.; Roger, A.; Sirdey, R. Towards better availability and accountability for iot updates by means of a blockchain. In Proceedings of the 2017 IEEE European Symposium on Security and Privacy Workshops (EuroS&PW), Paris, France, 26–28 April 2017; pp. 50–58.

160. Samaniego, M.; Deters, R. Blockchain as a Service for IoT. In Proceedings of the 2016 IEEE International Conference on Internet of Things (iThings) and IEEE Green Computing and Communications (GreenCom) and IEEE Cyber, Physical and Social Computing (CPSCom) and IEEE Smart Data (SmartData), Chengdu, China, 15–18 December 2016; pp. 433–436.

161. Zhu, C.; Leung, V.C.M.; Shu, L.; Ngai, E.C. Green Internet of Things for Smart World. *IEEE Access* **2015**, *3*, 2151–2162. [CrossRef]

162. Abedin, S.F.; Alam, M.G.R.; Haw, R.; Hong, C.S. A system model for energy efficient green-IoT network. In Proceedings of the 2015 International Conference on Information Networking (ICOIN), Siem Reap, Cambodia, 12–14 January 2015; pp. 177–182.

163. Nguyen, D.; Dow, C.; Hwang, S. An Efficient Traffic Congestion Monitoring System on Internet of Vehicles. *Wirel. Commun. Mob. Comput.* **2018**, *2018*. [CrossRef]

164. Namboodiri, V.; Gao, L. Energy-Aware Tag Anticollision Protocols for RFID Systems. *IEEE Trans. Mob. Comput.* **2010**, *9*, 44–59. [CrossRef]

165. Li, T.; Wu, S.S.; Chen, S.; Yang, M.C.K. Generalized Energy-Efficient Algorithms for the RFID Estimation Problem. *IEEE/ACM Trans. Netw.* **2012**, *20*, 1978–1990. [CrossRef]

166. Xu, X.; Gu, L.; Wang, J.; Xing, G.; Cheung, S. Read More with Less: An Adaptive Approach to Energy-Efficient RFID Systems. *IEEE J. Sel. Areas Commun.* **2011**, *29*, 1684–1697. [CrossRef]

167. Klair, D.K.; Chin, K.; Raad, R. A Survey and Tutorial of RFID Anti-Collision Protocols. *IEEE Commun. Surv. Tutor.* **2010**, *12*, 400–421. [CrossRef]

168. Lee, C.; Kim, D.; Kim, J. An Energy Efficient Active RFID Protocol to Avoid Overhearing Problem. *IEEE Sens. J.* **2014**, *14*, 15–24. [CrossRef]

MDPI
St. Alban-Anlage 66
4052 Basel
Switzerland
Tel. +41 61 683 77 34
Fax +41 61 302 89 18
www.mdpi.com

Energies Editorial Office
E-mail: energies@mdpi.com
www.mdpi.com/journal/energies